MANAGING DESTINATIONS

BRIDGING TOURISM THEORY AND PRACTICE

Series Editors:

Jafar Jafari
Department of Hospitality and Tourism, University of Wisconsin-Stout, USA.
Jafari@uwstout.edu

Noel Scott
School of Tourism, Edith Cowan University, Australia.
dr.noel.scott@gmail.com

Recognizing the increasing gap between what is researched in academic community and what is practiced in industry, this series aims to bring together academic and industry leaders in their respective fields to discuss, exchange, and debate issues critical to the advancement of tourism. The book series intends to not only create a platform for academics and practitioners to share theories and practices with each other, but more importantly to serve as a collaborative venue for meaningful synthesis.

Each volume will feature a distinct theme by focusing on a current or upcoming niche or "hot" topic. It shows how theories and practices inform each other; how both have evolved, advanced, and been applied; and how industry best practices have benefited from, and contributed to, theoretical developments. Volume editors have both strong academic credentials and significant consulting or other industry engagement experiences. Chapter contributors will be identified through professional conferences and trade conventions. In general, the book series seeks a synergy of how concepts can inform actions, and vice versa. The book series will inspire a new generation of researchers who can translate academic discoveries to deliverable results valuable to practitioners.

BRIDGING TOURISM THEORY AND PRACTICE
Volume 14

MANAGING DESTINATIONS: FROM THEORIES TO PRACTICES

EDITED BY

NOEL SCOTT
Edith Cowan University, Australia

MANUELA GUERREIRO
University of Algarve, Portugal

And

PATRÍCIA PINTO
University of Algarve, Portugal

emerald PUBLISHING

United Kingdom – North America – Japan
India – Malaysia – China

Emerald Publishing Limited
Emerald Publishing, Floor 5, Northspring, 21-23 Wellington Street, Leeds LS1 4DL

First edition 2024

British Library Cataloguing in Publication Data
A catalogue record for this book is available from the British Library

ISBN: 978-1-83797-177-0 (Print)
ISBN: 978-1-83797-176-3 (Online)
ISBN: 978-1-83797-178-7 (Epub)

ISSN: 2042-1443 (Series)

Printed and bound by CPI Group (UK) Ltd, Croydon, CR0 4YY

INVESTOR IN PEOPLE

CONTENTS

DESTINATION STAKEHOLDERS

INFRASTRUCTURE

RESEARCH AND KNOWLEDGE MANAGEMENT

LIST OF FIGURES

LIST OF TABLES

ABOUT THE CONTRIBUTORS

Juan José Albendín-Moya, PhD in Marketing, is a Lecturer at the University of Huelva (Spain) since 1993, in its Business Administration and Marketing Department. In recent years, he has developed his research activity through collaboration in a number of projects related to tourism and the application of new technologies to the marketing function of businesses. He is currently a part of the "Espomar" Project Working Group, in charge of studying the suitability, acceptability, and feasibility of potential maritime lines between the provinces of Cádiz and Huelva, in Spain, and the Algarve region, in Portugal.

Ali AlGassim, PhD in Tourism and Hospitality from Griffith University, Australia in 2015, is an Assistant Professor in Tourism and Hospitality Department at Umm AlQura University in Saudi. He has publications in peer reviewed journals in Tourism, Leisure and Hospitality Management domain. He has worked with many public and private companies in the sector. He recently worked on developing the historical village of Al-Juhfa for tourists, and published number of journal articles in tourism development. He is a member of Experts Program for Qualifying Specialists in UNESCO Conventions and Programs.

Diana Baus is enrolled in a PhD in Economics at the University of Innsbruck, Austria. She holds a Master's degree from the City University of New York. She works as an Associate Consultant at the Institute for Tourism, Zagreb. Her research interests are in macroeconomic aspects of tourism, global value chains, development and application of circular economy in tourism, and transfer of new technologies. She has been involved in a number of Interreg Europe projects for the European Union and the Tourism Action Plan/Event Tourism and Cultural Tourism for the Split-Dalmatia County Tourist Board.

Alexandra Bec, University of the Sunshine Coast, Australia, received her PhD from Southern Cross University in 2016. She is a skilled researcher and data analyst, with expertise in the areas of economic development, resilience, regional community development, community and stakeholder engagement, policy and planning, and change management. She has worked in industry, academia, and local government on research projects to inform regional community and economic development. She has taught both undergraduate and postgraduate courses, as well as supervised doctoral research candidates.

Monica Bernardi, Urban Sociologist with a PhD in Quality of Life in the Information Society, is Associate Professor at the Department of Sociology and

Social Research of the Milano-Bicocca University where she teaches Urban Tourism and Environmental Sociology. Her research interests focus on smart and sharing cities, practices of social innovation, platform economy, and urban tourism. She has been a visiting scholar at the GRI-Gyeonggi Research Institute, Seoul (2015); at TUFTS University, Medford, Boston (MA) (2017); and at the Dimmons Research Group, Open University of Catalunya, Barcelona (2019).

Nunzia Borrelli is an Associate Professor at University of Milan-Bicocca. In 2005, she obtained PhD in Spatial Planning and Local Development, at Turin Polytechnic. Since completing PhD, she has both taught and been involved in various research activities. She conducted field research projects pertaining to local development, ecomuseums, cultural heritage, and governance processes in Italy, the UK, USA, and China. She was a visiting as Fulbright scholar at Loyola University of Chicago at Portland State University, at University of Newcastle upon Tyne, and at Xiamen University (China). She has published several papers and three books.

Ana Cláudia Campos holds a PhD in Tourism, a MA in Tourism Management, and a BA in Philosophy. Currently, she is an Assistant Researcher at Research Centre for Tourism, Sustainability and Well-being (CinTurs), University of Algarve, Portugal. Her research interests are in tourism marketing, tourist experience, tourist psychology, experience design, co-creation, storytelling, and living labs and innovation in tourism.

Guilherme Castela, PhD in Applied Multivariate Statistics, Economics Degree, Masters of Quantitative Methods applied to Economics and Management and Advanced Studies in Applied Statistics, is a College Professor since 1989, focusing on economics, statistics, business strategy, game theory, and data analysis. He is an Assistant Professor at the Faculty of Economics, University of Algarve, Portugal; Director of the MSC Program in Health Care Services Management, Faculty of Economics, University of Algarve; and Member of the Research Centre for Tourism, Sustainability and Well-being (CinTurs). His research interests include data analysis, decision-making support, health management, management, economics, and tourism.

Vladislav Chernov, PhD, is an Associate Professor and Lecturer at the Department of International Communications, Service and Tourism in Far Eastern State Transport University, Russia; Pacific National University, Russia. His research interests include sustainable tourism development, tourism economy, the historical development of tourism in Russia, cross-border tourism, Sino–Russian relationship. He has over 200 academic manuscripts published including 8 books and has been recognized as one of the TOP 100 most cited tourism researchers in the Russian Federation. He is a member of the International Tourism Academy and a representative of the National Union of the Hospitality Industry Research Centre in the Far East of Russia.

Maria João Custódio holds a PhD in Management from University of Exeter Business School (UK) and Specialization in Marketing from ISCTE (Portugal). He is a researcher at CiTUR – Centre for Tourism Research, Development and Innovation, and a Guest Professor at University of Algarve – School of Management, Hospitality and Tourism. His research interests focus on tourism, marketing, and management, particularly in the impact of events, image perception, sports tourism, and elderly tourism. He shows intense participation in EU and national fundamental and applied research projects, ERAMUS+ projects, and holds publications in international peer reviewed journals and books, being also reviewer of well-established journals and received two tourism research prizes.

Yoná da Silva Dalonso, Graduated in Tourism and Hospitality, Master in Communication Science, and PhD in Geography, is a Professor and Researcher specializing in Tourism. She is currently the Vice Rector for Extension and Community Relations at the University of the Region of Joinville - Univille, Brazil, and leader of the Research Group on Tourism and Territory at Univille. She is member of the Right to Food Observatory for Latin America and the Caribbean (ODA-LAC) of FAO. Her publications include academic journal articles and books about tourism, gastronomy, public policies, events, tourism planning, and regional planning.

Saúl Neves de Jesus, PhD, is a Full Professor at the University of Algarve and currently Vice Rector. He is the Director of the PhD in Psychology and was a Director, President of the Scientific Board and President of the Pedagogical Council of the Faculty of Human and Social Sciences. He was the President of CIEO, Research Center ranked "Very Good" by FCT, and was a doctoral and postdoctoral fellowships evaluator. He has supervised postdoctoral fellows more than 40 doctoral and 50 master's theses. He published scientific books, more than 200 scientific papers, and has won national and international distinctions.

Diswandi is a Lecturer in the Faculty of Economics and Business, University of Mataram, Lombok. He obtained his PhD in Economics from Murdoch University in 2018. His research interest includes environmental economics, natural resource management, payment for ecosystem services, tourism economics, ecotourism, and sustainable development. Since 2021, he is the Vice Dean for academic affairs in the Faculty of Economics and Business, University of Mataram.

Elham Falatoonitoosi, PhD, is a Sessional Lecturer in the School of Business and Creative Industries at University of the Sunshine Coast, Australia. Elham published 12 peer reviewed papers in Tourism and operational management journals over the past 6 years and her first book was awarded a Gold Medal in ISEBA, Malaysia 2016. Elham's current research focuses on using mixed methods to analyze the impacts of sustainable development on regional tourism destinations. She is particularly interested in understanding how environmentally friendly and sustainability practices can enhance people's prosperity.

Soraia Garcês, PhD, is an Assistant Professor at University of Madeira – Research Center for Regional and Local Studies, and a Member of the Research Centre for Tourism, Sustainability and Well-being – University of Algarve. She won a postdoc research fellow with the project "Tourist Wellbeing in Madeira Island." Her research interests are tourism, creativity, innovation, education, and positive psychology. She has published in SCOPUS and Thomson Reuters journals, and was a team member of European research projects.

Manuela Guerreiro holds a PhD in Economics and Management Sciences, and an MSc in Cultural Management. She is an Assistant Professor in the Faculty of Economics, University of Algarve (Portugal). She is the Vice President of the Research Centre for Tourism, Sustainability and Well-Being (CinTurs) since 2018, where she is also an integrated researcher. She is the Program Leader of the Master of Marketing Management. Her current research interests include marketing and design, consumer behavior, consumer experiences, social media marketing, arts and events marketing, destination branding and destination image. She has published several articles in scientific peer reviewed journals, books, and international book chapters. She is also a coordinator or a participant in funded research projects in the fields of Tourism Management.

Damir Krešić has a doctorate from the Faculty of Economic and Business, University of Zagreb. He is the Managing Director, Institute for Tourism, Croatia, and Faculty of Economics and Business, Zagreb University, Croatia. His research interests include destination management and marketing, strategic planning in tourism, competitiveness and attractiveness of tourism destination, tourism sustainable and development, and application of ICT in tourism. He has completed numerous research projects in Croatia.

Júlia Maria Lourenço (*In Memoriam*) is a Civil Engineer from the Faculty of Engineering (University of Porto – FEUP); Postgraduate in Population and Development (ISS); PhD in Civil Engineering (Technical University of Lisbon – UTL); and a professor at the University of Minho. She conducted research and publications project mainly in the areas such as tourism planning and sustainable development policies for urban communities.

Ezio Marra is a Senior Professor of Environment Sociology at University of Milano-Bicocca. He is a Full Professor of Environment Sociology (University of Calabria 1994–2007, and University of Milano-Bicocca 2007–2018), Associate professor of Methodology of social research and Research and data processing techniques (University of Turin 1978–1994), Director of the degree course in Tourism management (2008–2015), Life member of ISA (International Sociological Association), Director of the Board of the Research Committee in «Urban Sociology» of the Italian Sociological Association – AIS (2001–2004), Director of the Computing Centre of the University of Calabria (1996–2000), and Director of the International Doctorate in "Sustainable Human Development" (2017–2018). His applied research is concerned with Sustainable Urban Tourism and Policies issues in metropolitan areas.

Teresa Medeiros is a Full Professor of Psychology at the School for Social Sciences and Humanities, University of the Azores, Portugal. She obtained her PhD in Education Sciences – specialization in Educational Psychology, at the University of Coimbra. She is a researcher at the Institute for Cognitive Psychology and Social and Human Development at the University of Coimbra's School of Psychology (IPCDSH). Her interest in topics related to psycho-gerontology, development psychology, educational psychology, senior tourism, and well-being has motivated her to conduct research and projects in this field.

José Mendes holds a PhD in Psychology (Health Psychology) and a PhD student in tourism at the IGOT-UL, Portugal. He is affiliated to INTELECTO – Psicologia & Investigação, Ponta Delgada, and to IPCDSH, University of Coimbra. His areas of interest in research include body image, aging, senior tourism, health and well-being in tourism, and ICT in tourist behavior.

Ana Isabel Moniz holds a PhD in Economic and Business Sciences. She is an Assistant Professor and the Head of Management Department at the School of Business and Economics, University of the Azores, Portugal, where formerly she was the Head of the BSc in Tourism. She is a founding member and a researcher at the Centre of Applied Economics Studies of the Atlantic. Her research interests and publications focus on sustainable tourism, destination image, tourism planning, senior tourism, consumer behavior, customer satisfaction, impulse buying behavior, and services marketing.

Miriam Mota is a PhD student in Tourism at the Faculty of Economics at the University of Algarve (Portugal). She is a member of the Doctoral Student Network at the Research Centre for Tourism, Sustainability and Well-Being (CinTurs). She has a Master's Degree in Tourism Business Management from the State University of Ceará (Brazil) and a Bachelor's Degree in Tourism from the Federal University of Maranhão (UFMA). Her research interests are in the areas of heritage tourism, cultural studies, destination branding and destination image.

Vanessa Oliveira, Master in Marketing and Bachelor in Tourism from the University of Algarve, is a Collaborator Researcher at CiTUR – Centre for Tourism Research, Development and Innovation, and Superior Technician of the Evaluation and Quality Office of the University of Algarve. Her research is focused on the themes of tourism, events impact, sustainability, and local communities. She participates in several applied research projects related with events impact monitoring, being an expert in the data organization and administration of fieldwork/surveys in tourism destinations to visitants and residents.

Alexandre Panosso Netto, graduated in Tourism and Philosophy, master's in History, PhD in Communication Sciences, and Post-doctorate in Tourism from Universidad Europea Miguel de Cervantes (Spain), is a Full professor at the School of Arts, Sciences and Humanities at the University of São Paulo, Brazil. His publications include academic journal papers and books about tourism, education, public policies, events, tourism planning, regional planning, and epistemology of tourism. He is the president of the Brazilian National Association for Research and Graduate Studies in Tourism-ANPTUR (2020–2020). He has worked with Tourism Higher Education Institutions in several Latin American countries.

Luís Nobre Pereira is a Professor at the School of Management, Hospitality and Tourism of the University of Algarve. He is the Vice president of the Research Centre for Tourism, Sustainability and Well-being (CinTurs) since 2018. He is active in conducting funded research projects in the fields of Tourism and Hospitality Management. Professor Luis Pereira holds a PhD degree in Quantitative Methods Applied to Economics and Management. His research interests include tourism demand modeling and forecasting, decision support systems for tourism, sustainable tourism, revenue management, and dynamic pricing for the hospitality industry.

Mariia Perelygina, PhD, is currently a Postdoctoral Researcher at the School of Hospitality Management and Tourism, Technological University Dublin in Ireland. Recently, she has obtained PhD in Hotel and Tourism Management from the School of Hotel and Tourism Management, The Hong Kong Polytechnic University. In 2010, she obtained her Specialist's degree in Tourism and Hospitality Management from Udmurt State University (Izhevsk, Russia). In addition to teaching and research experience, Mariia has diverse practical experience in the travel industry. Her research interests include cross-cultural management in tourism, digitalization and smart tourism, strategic management in tourism.

Fernando Perna, PhD in Applied Economics, is a Full Professor at the University of Algarve – School of Management, Hospitality and Tourism, Researcher and National Director of CiTUR – Center for Tourism Research, Development and Innovation. His research interests focus on tourism, regional development, sustainability, impact of events, coastal zone management, and blue economy. He participates and coordinates national fundamental and applied research projects, and international research and educational projects in Europe, Africa, and Asia. He has publications in peer reviewed journals in Tourism, Leisure and Hospitality Management domain. He is an expert member of National Committee of Recreational Boating of Portuguese XXI Government Sea Ministry.

Patrícia Susana Lopes Guerrilha dos Santos Pinto holds a PhD in Quantitative Methods Applied to Economics and Business. She is an Associate Professor at the Faculty of Economics, University of Algarve, Portugal. She is the coordinator of the Research Centre for Tourism, Sustainability and Well-being (CinTurs) and member of the directive board of the PhD in Tourism of the University of Algarve. Her current research interests are in applied statistics and modeling especially in Tourism Management (Sustainable Tourism, Destination image and branding, Marketing segmentation, Consumer behavior). She is author or coauthor of about 90 scientific papers in journals indexed in Scopus or Web of Science. She is also a coordinator or a participant in funded research projects in the fields of Tourism Management.

Margarida Pocinho, PhD, is an Associate Professor with Aggregation at Madeira University and is a Member of the Research Centre for Tourism, Sustainability and Well-being, University of Algarve. She is evaluator of project fellowships (FCT and H2020) and was a Fullbright fellow from New York University. Her research interests are tourism, creativity, innovation, education, and positive psychology. She has published in SCOPUS and Thomson Reuters journals and has supervised a large number of MA and PhD dissertations. She was Co-Guest Editor of indexed journals and is team leader of European Projects.

Paula Cristina Almeida Remoaldo is a Full Professor at the Department of Geography of the University of Minho (Portugal) and holds a PhD in Human Geography (with Louvain-La-Neuve University collaboration). She was the Director of the Landscape, Heritage and Territory Laboratory, 2017–2020, Director of the Department of Geography (2000–2019), and Director of the Master of Geography (2013–2015) and of the PhD in Geography (2015–2017). She has 31 years of independent research in the demographic, social, and cultural analysis of heritage in small cities and rural territories. She conducts research mainly in the area of cultural and creative tourism and local/regional development.

Elisa Rossi is a student at University of Milan-Bicocca. Graduated in Science of tourism, she is currently specializing in Economy of tourism. During her studies she focalizes on tourism, in particular on the National Strategy of Internal Areas (SNAI). Her research interests include tourism in internal areas of Italy and the application of sustainability to the tourism sector.

Edson Santos is a PhD candidate in Tourism at the Faculty of Economics of the University of Algarve, a Research Member at the Research Centre for Tourism, Sustainability and Well-being (CinTurs), and a faculty member at the School of Business and Governance of the University of Cape Verde. His line of study includes destination management and marketing, sustainable tourism development, small island tourism, resident perception of tourism development, and support for tourism development.

Maria Margarida Santos, PhD in Tourism, is a Senior Lecturer and Head of the Marketing Department at the School of Management, Hospitality and Tourism of the University of Algarve and Researcher of the Research Centre for Tourism, Sustainability and Well-being (CinTurs). Her research lines include sustainability in tourism and hospitality, innovation in tourism, senior tourism, digital marketing, and gastronomy in tourism. She participates in funded Erasmus+ and research projects in tourism and hospitality and published in international peer reviewed journals and books. She is also Editor of the journal Tourism & Management Studies.

Akhmad Saufi, PhD in International Tourism and Hospitality Management from Griffith University, is a Senior Lecturer of Tourism Management in the Department of Management at the Faculty of Economic and Business University of Mataram,

Chairman of the Monitoring Center of Sustainable Tourism Observatory of Lombok, Indonesia, member of Indonesia Sustainable Tourism Council, Master Assessor and Auditor Training Program for Sustainable Tourism Certification, and has publications in peer reviewed journals in the domain of tourism, leisure, and hospitality management. His research interests are in the areas of host community participation, sustainable practices in tourist destinations, tourism entrepreneurship, halal tourism, heritage, and cultural tourism.

Noel Scott, PhD, is an Adjunct Professor of Tourism Management, in the Sustainable Research Centre, University of the Sunshine Coast, Queensland, Australia. His research interests include the study of tourism experiences, destination management, and marketing. He is a frequent speaker at international academic and industry conferences. He has over 300 academic articles published including 16 books. He has supervised 23 doctoral students to successful completion of their theses. He is on the Editorial Board of 10 journals, a member of the International Association of China Tourism Scholars, and a Fellow of the Council for Australasian Tourism and Hospitality Education.

Bernardete Sequeira holds a PhD in Sociology. She is an Assistant Professor in the Faculty of Economics, University of Algarve (Portugal). She is a member of the Board of the Masters in Sociology and subdirectory of the Degree in Sociology at this institution, and an integrated member of the Interdisciplinary Centre of Social Sciences – CICS.NOVA. She is also a collaborator researcher of CinTurs – Research Centre for Tourism, Sustainability and Well-being. She has published articles in scientific journals and international book chapters. Her current research interests include sociology of organizations and work, sociology of knowledge, knowledge management and tourism.

João Albino Silva, PhD in Economics, is a Full Professor at the Faculty of Economics, University of Algarve, Portugal; Director of the PhD Program in Tourism; Head of the Economics Department, Faculty of Economics, University of Algarve; President of the Scientific Committee of the Faculty of Economics; Scientific Coordinator of CinTurs – Research Centre for Tourism, Sustainability and Well-being; President of the external evaluation committees (A3ES) of Tourism graduation and post-graduation; and Scientific coordinator of the Algarve Sustainable Tourism Observatory. His research interests include tourism planning, management and development, quality and management of tourism destinations, economics of tourism and environment.

Osvaldo Silva is an Assistant Professor at the Faculty of Sciences and Technology of the University of the Azores, Portugal, and a researcher at the Interdisciplinary Centre of Social Sciences of the University of the Azores and New University of Lisbon, CICS.UAc/CICS.NOVA.UAc. He holds a PhD in Mathematics – specialization in Probability and Statistics. His research interests and publications focus on multivariate data analysis and applied statistical methodologies, namely in social sciences (senior tourist, ICT in tourism, consumer behavior, human resource management).

Lorenza Sganzetta is a geographer, urban and sustainable food systems planner, researcher on the themes of urban regeneration, environmental sustainability, and circular economy. She worked as a consultant for FAO, IOM, and the Feltrinelli Foundation, as a fellow and postdoctoral fellow for Milan Polytechnic, Bicocca University and Scuola Superiore Sant'Anna, contributing to planning, participatory design and cartographic project analysis. Her research interests are sustainable urban planning, sustainable food systems, and circular economy.

Liubov Skavronskaya, PhD, is a Research Assistant and Sessional Lecturer at the School of Business and Creative Industries, University of the Sunshine Coast (USC), Australia. Prior to coming to USC, Liubov worked as a Research Assistant at Griffith University, Australia, and a Lecturer at Shunde Polytechnic University in China. She also graduated with a Master of Tourism Management from South China University of Technology and Bachelor (Hons) from Far East State Transport University in Russia. Liubov conducts research in English, Mandarin, and Russian languages. Her interests include emotions, memorable experiences, tourism experience design, cognitive psychology, and China tourism.

Olga Tzatzadaki is a PhD candidate in Urban Planning and Public Policy in the University IUAV of Venice. Her research interests focus on undertourism and overtourism (UNESCO: Venice-Mestre and the Dolomites of Belluno Province), as well as on the importance of narration and storytelling in urban planning documents and in literary works, as evaluation methods for urban planning practices. Her publications include academic journal articles and her work has been presented at national and international research conferences. She holds a BA in Urban Planning, a BA in Infrastructural Civil Engineering, and a Master's degree in Urban Planning.

Alfonso Vargas-Sánchez, PhD in Business Administration, is a Full Professor at the University of Huelva (Spain), with research interests focused on strategic management, particularly of tourist companies and destinations. He has carried out academic activities, both in researching and teaching, in different universities and higher education institutions around the world, in the five continents, and authored a number of books and articles published in top-ranked national and international scientific journals. He has been the promoter and Editor-in-Chief of "Enlightening Tourism: A Pathmaking Journal", listed in Scopus since 2018, during the period 2011–2022.

João Vidal, PhD in Tourism, PhD in Law, Law Degree, MBA in General Management, Lawyer, an Invited Assistant Professor at the Faculty of Economics, University of Algarve, Portugal, is a member of Research Centre for Tourism, Sustainability and Well-being (CinTurs); investigator in the R&D project called "Legal Treatment of Collaborative Tourism and Online Platforms," funded by the Spanish Government, housed at the Faculty of Law of the University of the Balearic Islands; and referee in Jurismat, a law scientific journal of Lusófona University, Portugal. His research interests include tourism management, tourism law, civil liability in tourism, and decision-making support.

ACKNOWLEDGMENT

This work was supported by the National Funds provided by FCT – Foundation for Science and Technology through project UIDB/04020/2020.

Chapter 1

INTRODUCTION

Noel Scott
Edith Cowan University, Australia

Manuela Guerreiro
University of Algarve, Portugal

Patrícia Pinto
University of Algarve, Portugal

This book draws upon papers submitted to the t-Forum 2020 conference entitled "Breaking Barriers in the Changing World: Bridging Theory and Practice in Tourism" held in Algarve, Portugal (18–20 March 2020). The t-Forum is about transfer of knowledge/intelligence to and within tourism. The intent is to couple concepts and their applications to bring together researchers and practitioners who work on tourism from a variety of approaches, disciplines, and fields. The papers from the conference were supplemented by a general call for chapters. The chapters in this edited book are all examples of studies that seek to transfer intelligence to and within tourism. The book may be especially of interest to those involved in the management of tourism destinations. The chapters have been divided into five sections: strategy, new markets, stakeholders, infrastructure, and research and knowledge management. Within these sections, cross-cutting themes include the importance of social and environmental sustainability and integration of governance.

The first section of destination strategy discusses three cases from Italy and one from Far East Russia. Chapter 2 examines the implementation of a national strategy for the development of rural Italian regions and its relationship with tourism. The case illustrates the difficulties of aligning

Managing Destinations: From Theories to Practices
Bridging Tourism Theory and Practice, Volume 14, 1–3

ISSN: 2042-1443/doi:10.1108/S2042-144320240000014001

development strategy across levels of government and with industry and the community. Chapter 3 contrasts and compares the development of Genoa, Turin, and Milan over the past decades and their movement from an industrial to a postindustrial experience economy. Chapter 4 discusses the development of tourism in the Russian Far East focusing on the "Twin Cities" of Blagoveshchensk (Russia) and Heihe (China). This is discussed as an example of an experiential core developing at a peripheral destination. The effects of politics between China and Russia on cross-border tourism are examined, and the recent effects of COVID-19 are noted. Chapter 5 returns to Italy and the urban development strategy of Mestre on the outskirts of Venice. The chapter contrasts undertourism in Mestre, with overtourism suffered by Venice. The authors present a well-argued case for the expansion of the core tourism area of Venice to include Mestre, and in doing so highlight the politics of place development.

The section on new markets has two chapters from small islands in the Atlantic Ocean. Chapter 6 provides the results of a tourist well-being project in Madeira Island, Portugal. This study allows for design of customized experiences that can enhance tourists' psychological profile and their well-being. These experiences include gastronomy, walks along irrigation channels (Levadas), and scenic landscape. Chapter 7 presents results of research that segments and characterizes senior tourists who used online information to book their trip to the Azores Region in Portugal. This highlights the importance of information and communication technology and its use in understanding tourist behavior.

The third section focuses on destination stakeholders. In Chapter 8, the authors provide a case study of the development of a small village in Saudi Arabia based on its archaeological heritage and religious significance. This raises questions concerning the involvement of local residents in tourism development and how they can benefit from it. Chapter 9 presents the intertwining model which examines the level of influence of events on the public policies in a destination and its interface with the different actors involved. The chapter identifies stages in the evolution of public policies and their relationship to the networks of different actors, noting that stakeholders can have multiple roles. Chapter 10 examines support by residents of Santiago Island, Cape Verde, for tourism during COVID-19. The results indicate that residents' support for tourism was positively influenced by their quality of life. Thus, the more tourism benefits locals, the more they support tourism. Chapter 11 reports on stakeholder views of the historic centre of São Luís, Brazil, a UNESCO World Heritage site. While World Heritage Site listing protects a site, it can improve tourism but bring little benefit to residents. This study examines the perceptions of residents and other stakeholders of this site.

The fourth section discusses infrastructure and its importance and development for tourism. Chapter 12 discusses energy, specifically the adoption of solar energy in Croatia. This is done by examining good practices in several locations around the Mediterranean Sea. The impact of politics and public opinion on solar power strategy are noted. Chapter 13 discusses the legal infrastructure of tourism and the use of arbitration from the perspective of litigation. Tourism disputes often involve two legal systems, and organizations can use arbitration instead of courts to get a good effect. The next chapter examines ports and sea transport. Chapter 14 provides results from research as part of the Spain–Portugal Cross-Border Cooperation EU Program. The aim is to design a sustainable and transboundary maritime transport system in the Gulf of Cádiz, which includes the Spanish provinces of Cádiz (its Atlantic coast) and Huelva, together with the Portuguese region of the Algarve. The findings indicate demand for potential tourism packages created around boat trips with sun-beach, gastronomy-wine, and culture-landscape vectors as the main attractors.

The final section and chapter discuss the importance of research in destination management. While many of the chapters in previous sections use research of various types, Chapter 15 seeks to develop innovation in a destination using the concept of a living lab, which is a model of open innovation characterized by a networked approach to innovation, enabling local communities' participation in the sustainable development of tourism. This chapter is a good practice case targeted at transfer of knowledge/intelligence to and within tourism. Finally, Chapter 16 provides some key lessons for management of tourism destinations and directions for further research.

DESTINATION STRATEGY

Chapter 2

TOURISM IN PERIPHERAL AREAS
National Strategy for Italian Regions

Nunzia Borrelli
Univeristà di Milano Bicocca, Italy

Lorenza Sganzetta
Univeristà di Milano Bicocca, Italy

Elisa Rossi
Univeristà di Milano Bicocca, Italy

Abstract: This chapter discusses the development of tourism in peripheral areas, prompted by the "National Strategy for Inner Areas" developed by the Italian National Government. The strategy, backed by policymakers, business owners, local communities, and environmental nongovernmental organizations (NGOs), aims to stop demographic decline by boosting sustainable tourism practices. A case study of Valle D'aosta examines the problems and their solution in the implementation of the strategy. It discusses how to make strategy implementation less complex and whether sustainable tourism is possible in such areas – whether it is an oxymoron, or whether it is a utopia that is worth pursuing. **Keywords**: Sustainability; peripheral regions; inner areas; local development practices; spatial policies

INTRODUCTION

A large part of Italy, accounting for more than half of the territory and a little less than a quarter of the population, is composed of areas characterized by marginalization, degradation, and abandonment and is located far from

Managing Destinations: From Theories to Practices
Bridging Tourism Theory and Practice, Volume 14, 7–21

ISSN: 2042-1443/doi:10.1108/S2042-144320240000014002

the big cities. Such areas in peripheral regions, called "inner areas," have undergone a progressive marginalization and de-anthropization, with a reduction in the use of territorial capital. Their small, inhabited centers offer limited access to essential service (Agenzia per la Coesione Territoriale, 2013; Lucatelli, 2016; Mantino & Lucatelli, 2016). While they have many problems and difficulties, still they are equipped with precious resources. Recently, the national government initiated tourism-focused development to enhance the local resources located in these areas.

Strengthening inner areas through tourism implies changing preconceptions about development and providing "governance capacity" in these locations (Borrelli, 2009). The goal is to implement "active landscape conservation" (Salvatore, 2015) based on the principle of coevolution among place, community, and tourism economy. This means opting for sustainable practices that aim to reactivate the place's value and the local community. The Italian government has taken the potential of inner areas into account in the National Reform Program developed by Dipartimento per lo Sviluppo e la Coesione Economica (2013), which has adopted the National Strategy for Inner Areas, aiming to stop the demographic decline and boost sustainable tourism practices.

This chapter discusses the most important steps in the development of the concept of sustainability related to tourism and its connection with regional development. It presents the process of constructing the national strategy for the inner areas, identifying the actors involved, and clarifying the strategies needed to attain sustainability and sustainable tourism. A case study helps to suggest the strengths and weaknesses of the national strategy implementation and of sustainable tourism.

DEVELOPING SUSTAINABILITY IN TOURISM

The first signs of interest toward a more respectful approach to local culture and the environment – crucial elements of the tourism product – can be found in the Manila Declaration on World Tourism that resulted from the 1980 World Tourism Conference held in Manila:

> Tourism resources available in the various countries consist at the same time of space, facilities, and values. These are resourcing whose use cannot be left uncontrolled without running the risk of their deterioration, or even their destruction. The satisfaction of tourism requirements must not be prejudicial to the social and economic interests of the population in tourist areas, to the environment or to natural resources, which are the fundamental attraction of tourism, and historical and cultural sites. All tourism resources are part of the heritage of humanity. National communities and the entire

> international community must take the necessary steps to ensure their preservation. The conservation of historical, cultural, and religious sites represents at all times, and notably in time of conflict, one of the fundamental responsibilities of States ... International cooperation in the field of tourism is an endeavour in which the characteristics of peoples and basic interests of individual States must be respected. (WTO, 1989, p. 3)

In September 1985, the United Nations World Tourism Organization (UNWTO) General Assembly held a meeting in Sofia, Bulgaria, during which the Tourism Bill of Rights and Tourist Code was drawn up. This bill acknowledges the importance of tourism and its positive effects on an economic, social, and cultural scale, while also recognizing its role in "improving mutual understanding, bringing peoples closer together and, consequently, strengthening international cooperation" (WTO, 1985, p. 1). Because of this, it was deemed necessary to establish the rights and duties of the individual states and the host communities, as well as of tourists and professionals in this industry.

Based on the concept of sustainable development defined at the Brundtland Report (1987), the WTO developed its definition of sustainable tourism: "Tourism that takes full account of its current and future economic, social, and environmental impacts, addressing the needs of visitors, the industry, the environment, and host communities" (UNWTO & UNEP, 2005, p. 24). In March 1990, the Globe '90 Conference organized in Canada was "identified as the first international attempt to incorporate the concept of sustainable development into that of sustainable tourism" (Borrelli & Bernardi, 2019, p. 34).

The turning point came in 1995 when the Charter for Sustainable Tourism was developed at the first World Conference on Sustainable Tourism in Lanzarote, Spain. This defines eighteen fundamental principles that tourism must follow to be considered a sustainable activity. The key points are respect for the environment and local cultures, equal distribution of the economic benefits that derive from tourism, and cooperation on a local, national, and international level. The charter also provides concrete lines of action that countries and private operators in the tourism industry should undertake, as well as encouraging tourists themselves to adopt patterns of consumption that respect the environment and the host community.

In 1996, UNWTO, together with the World Travel and Tourism Council and the Earth Council, published Agenda 21 for travel and tourism: toward environmentally sustainable development (UNWTO, 1996). This delineates, following the principles adopted after the 1992 Rio Summit, practical measures that governments and private businesses can adopt to make the industry more sustainable. It recognizes the importance of protecting natural and cultural resources for tourism, given that these constitute the core of its activities. It highlights the priority areas of intervention, defining objectives, and suggesting measures to adopt for achieving sustainable tourism.

In the same year, the Bureau International du Tourisme Social held the Montreal Conference, during which the Montreal Declaration was written, and an attempt was made to understand the problems and the prospects of social tourism, a tourism model that is nondiscriminatory and aims at integration. Therefore, this became the basis on which to build sustainable tourism since it lends itself to social cohesion (Montreal Declaration, 1996).

In 1997, three important international meetings took place on the subject of tourism, which in turn produced declarations of their own. One, International Conference of Environment Ministers on Biodiversity and Tourism, held in Berlin in 1997, produced the Declaration on Biological Diversity and Sustainable Tourism. Two, First International Conference in Calvià led to the Declaration on Tourism and Sustainable Development in the Mediterranean. Three, Manila Declaration on the Social Impact of Tourism was developed and promoted by the UNWTO.

In April 1999, the United Nations Commission on Sustainable Development approved a report entitled Tourism and Sustainable Development (Decision 7/3). This states that travel and tourism can offer economic, ecological, and social development because it has less impact on natural resources and the environment compared to most other industries; moreover, local culture and heritage and the natural environment are what motivate travel and, therefore, the tourism is incentivized to protect these assets. In 2001, the principles of the Lanzarote Charter were reviewed and updated in the Rimini Charter, resulting from the Second International Conference for Sustainable Tourism during which particular attention was given to mass tourism destinations in coastal areas and fragile environments.

The year 2002 was important for sustainable tourism as it was designated as the International Year of Ecotourism. In the same year, Quebec City hosted the World Ecotourism Summit, supported by UNWTO and UNEP, during which guidelines and methodologies were provided for the development of ecotourism. Still in the same year, the international community met in Johannesburg for the 2nd World Summit on Sustainable Development, during which the program for Sustainable Tourism – Eliminating Poverty was presented. Between 2002 and 2016, several other events and documents were promoted and produced to reinforce the relevance of sustainability in tourism (Figure 1).

This historical description is useful to underline some key points. Sustainable tourism identifies, as the center of its strategies, the activation of the local community and the construction of multilevel actor networks to guide the tourism planning processes.

In other words, the affirmation of the sustainability principles in tourism has changed its planning processes by expanding the network of actors involved and giving space not only to economic actors closely linked to its sphere but also to actors who deal with climate change, gender equality, social inclusion, and other issues connected to sustainability. As a result, the

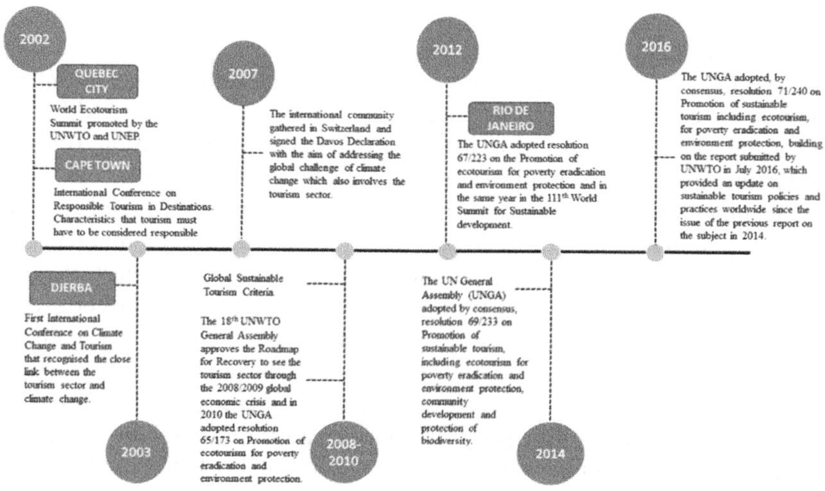

Figure 1. Sustainable Tourism Timeline Source.

main goal of tourism strategies is to generate economic wealth, enhance the role of the local community, and protect the environment. Rather than being damaged, the environment should be valorized by tourism initiatives.

Fragility or Potential for the Territory

Most of the Italian territory is characterized by small population centers with limited access to essential services. These areas, defined as "inner" in relation to large and medium urban areas and their connecting networks, are often characterized by natural (forests, protected areas, agricultural, and agri-food production) and cultural (archeological heritage, abbeys, small museums, trade centers) capital. Since the 1950s, a substantial part of these inner areas has undergone a process of marginalization and de-anthropization (Lucatelli & Carlucci, 2013).

The Department for Economic Development and Cohesion of the Ministry of Economic Development has drafted a national plan for the development of the inner areas of the country in integrated and uniform planning. This plan, "Strategia Nazionale Aree Interne," defines an innovative national policy of development and territorial cohesion to counter the marginalization and the phenomena of demographic decline typical of the internal areas of the country, treating as inner areas those significantly distant from the centers offering essential services (education, health, and mobility), based on an accessibility indicator calculated in terms of minutes of travel with respect to the nearest service hub. Concerning the territory classified as internal areas, the strategy distinguishes between two modes of local development: intensive and extensive development. The former refers to all those changes that increase the per capita well-being of the residents without increasing production. Extensive development refers to all those changes that in addition to increasing the well-being also increase the scale of production processes (Dipartimento per lo Sviluppo e la Coesione Economica, 2013).

The inner areas are in turn divided into intermediate, peripheral, and ultra-peripheral areas. These represent about 53% of the Italian municipalities (4,261) to which 23% of the population belongs (Istat, 2011), just over 13.5 million inhabitants residing in a portion of the territory that exceeds 60% of the country (Barca, Casavola, & Lucatelli, 2014). Unlike the inner areas, municipalities are "poles" with a wide range of services, including a complete upper secondary school, at least one first-level hospital (containing a range of services: general medicine, general surgery, orthopedics and traumatology, cardiology therapy, laboratory, and imaging facilities), and at least one medium-small railway station. "Belt" municipalities are less than 20 minutes from the nearest pole while inner areas are over 20 minutes away (Table 1).

Most of the inhabitants of the inner areas (8.8 million people) live in the intermediate municipalities, 20–40 minutes away from the nearest pole. A further, 3.7 million live in peripheral municipalities, while another 670,000 people live in the outermost areas or municipalities, mostly on mountains or islands, at least 75 minutes away from the nearest center (Barca et al., 2014). The methodology adopted by the Ministry of Economic Development identifies the nature of an inner area in terms of the distance from the essential services. The degree of the periphery is identified by certain aspects (school, health, and rail transport services).

At the same time, the demographic changes of the last century which characterized the poles, belt areas, and inner areas and which have led to an increase in the elderly population (65+) in the peripheral and outermost areas, have contributed to the redefinition of social functions and territorial

Table 1. Inner Areas Classification.

Municipality Classification	Category	Distance from the Nearest Pole
Pole	Pole	–
Belt	Peri-urban areas	$t < 20$ minutes
Intermediate	Inner areas	20' $< t <$ 40 minutes
Peripheral	Inner areas	40' $< t <$ 75 minutes
Ultra-peripheral	Inner areas	$t > 75$ minutes

Source: Authors' own based document published by Dipartimento per lo sviluppo e la coesione economica (https://www.agenziacoesione.gov.it/wp-content/uploads/2021/01/Nota_metodologica_Aree_interne-2-1.pdf).

inland areas. There are even more visible consequences, such as the loss of active protection of the territory and the decrease in agricultural activities (Lucatelli & Carlucci, 2013).

These characteristics, although a symptom of territorial fragility, can be transformed into innovative development potential following the trajectories of environmental preservation and the capitalization of the social context in the area. A public policy must, therefore, aim at the renewal and promotion of the tools available in the inner areas, although apparently fragile, transforming them into resources for the improvement of the quality of life. For these reasons, this study seeks to understand the strengths and weaknesses in the implementation of a public policy to the inner areas (mainly via tourism) to make these areas more robust and their implications to the Italian society.

The Policymaking Process

The formulation of a national strategy for the development of the inner areas was started in September 2012 by the Minister for Cohesion, Fabrizio Barca, with the support of a Technical Committee for Inner Areas. After a phase of dialogue with the representatives of the various regions, a document relating to the Strategia Nazionale Aree Interne was drawn up and incorporated into the Partnership Agreement or Accordo di Partenariato (Dipartimento per lo Sviluppo e la Coesione Economica, 2013) between Italy and the European Commission. Accordo di Partenariato, is a document, specified by European Community Regulations with which each member state is called to define its strategy, priorities, and methods of employment of the European structural funds for the period 2014–2020. This document was prepared by Italy and approved by the EU Commission and defined strategies, methods, and priorities for spending resources cofinanced by the European Structural and Investment Funds (ESI Funds) for the 2014–2020 programing cycle.

Strengthening and preserving the inner areas received particular attention in the guidelines for the programing of the regional policy methods and objectives for Effective Use of Community Funds 2014–2020 and in the context of the same regional cohesion policy for the cycle 2014–2020.

In line with a place-based approach (Barca, 2009) of which this process can be considered an interesting implementation experiment, in the Strategia Nazionale Aree Interne (SNAI) Strategy for the Development of Inner Areas (Dipartimento per lo Sviluppo e la Coesione Economica, 2013), the definition of development strategies is based on an open comparison, through round table discussions, between stakeholders of central and subnational government and local communities. The idea is to systematize the potential of the area, in a phase of co-planning of interventions, aimed at changing the approach of the territories to local development and related policies.

Through co-planning, SNAI involved the local communities of the inner areas in initiatives aimed at promoting dialogue and meeting of the population with local and national administrations, as well as with external experts. The project aimed at contributing to the process of reorganization of public services in the territory of the inner areas of the country, ensuring the identification of effective management models that are consistent with the needs of the territories. To this end, starting from 2015, and once the partnership agreement with the European Commission was signed, the Italian government (through the Inner Areas Technical Committee) carried out intense joint work in selecting the territorial contexts of intervention of the strategy. As a result, 72 "project areas" were identified (Figure 2) distributed in all regions and in the autonomous province of Trento (Lucatelli & Storti, 2019). These project areas comprise more than 1,066 municipalities (13% of the municipalities Italians who occupy almost 17% of the national territory) and over two million inhabitants.

The regions selected, in agreement with the Ministry of Economic Development, became the focus of the policy intervention. To be selected, the municipalities had to belong to the classification of the inner areas as defined by the national mapping, with priority to peripheral and ultra-peripheral areas, and to present critical values in terms of demographic, economic, social, or environmental indicators. Preference was given to applications submitted which promoted the creation of Unions or Associations with a social or cultural purpose of effective services, have adequate capacity in the field of local development planning and created forms of partnership (Lucatelli, 2016).

Once the areas were identified, the second step consisted of the construction of the area strategy or "Strategia di Area." This is a plan for sustainable competitiveness aimed at addressing the demographic decline that characterizes these areas in the medium term. The strategy provides for integrated interventions/area projects to be defined considering both the aspects of local development promotion and those relating to the rebalancing of basic services for communities.

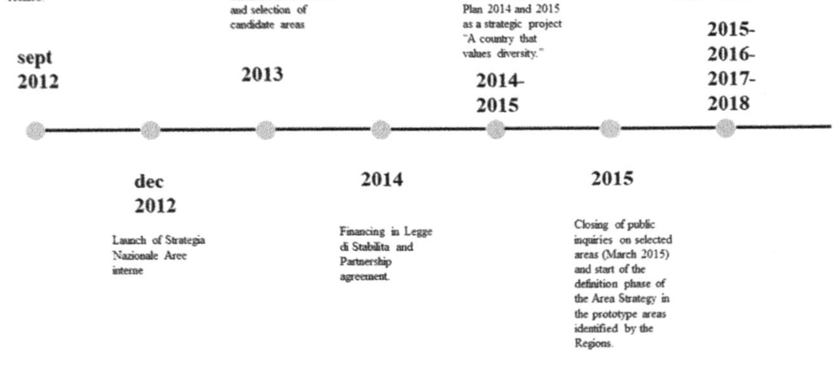

Figure 2. Strategia Nazionale Aree Interne Steps.

The strategic processes were developed in phases of increasing complexity, starting from the writing of a first document regarded as a "draft strategy" – a proposal for priority intervention and identification of a generic aspiration of the area (Lucatelli, 2016). The next phase saw the definition of a path that connected local development and services, starting from experiences and knowledge rooted in the territory (Lucatelli & Storti, 2019).

The Strategia di Area, once shared by the Inner Areas Technical Committee and the Region, was sent to the Territorial Cohesion Agency, which started the preparation phase of the fourth and final document: the so called "Accordo di Programma Quadro" – Framework Program Agreement (Dipartimento per lo Sviluppo e la Coesione Economica, 2014). As part of the negotiated programming, which allows autonomous regions and provinces to agree with the government on objectives, sectors, and areas in which to carry out interventions for the development of the regional territory. The framework program agreement defines, for a specific sector of intervention, works and financing, as well as procedures for monitoring the implementation of investments to be covered in the institutional arrangements of the state–region program. The latter is the instrument for implementing the institutional agreement between the project area, the region, and the state. This was an agreement which contained activities and interventions to be carried out, with the times and methods of implementation; the responsible subjects and their commitments; the necessary financial resources, out of public funds or obtained through private financing, and the procedures and subjects responsible for monitoring and verifying the results (Lucatelli & Storti, 2019).

The Role of Sustainability in the Strategy

As evidenced by the classification of the Ministry, the inner areas have undergone over the years a redefinition of their territorial human and social capital resources. This has determined, in part, a weakening of the economic–productive capacity of these areas which need to find a use that guarantees their sustainability over time. Thanks to the same analysis, it was possible to bring to light the potential of the landscape resources of these depopulated and fragile territories.

Currently, in the inner areas, the agricultural, pastoral, and forestry sectors almost always take on a leading role, both as the only economic driving force and for the value of environmental care and prevention. Agricultural land management plays a particularly key role in areas characterized by elevated levels of landslide risk and hydrogeological instability (Lucatelli & Carlucci, 2013). Farming in mountain areas contributes to their vitality and makes an essential contribution to the maintenance of biodiversity and to the fight against soil degradation. The ability to live and interpret the forest not as a mere effect of abandoning agriculture but as an opportunity can make it possible to achieve positive impacts on the environment and the creation of new jobs through improved forest management. Each of these activities contributes to enriching the attractive cultural fabric of these areas and preparing them for sustainable tourism activities that systematize local resources in the preservation of natural capital and the biodiversity of nature.

The SNAI strategy has already introduced the theme of rural development in future development plans, supporting the synergistic action of different sectors of public administration and different policies ranging from tourism to agriculture, to local development, to innovation, to education, mobility, and health. Of these sectors, the potentially most promising and effective one is tourism. In its sustainable and environmentally friendly version, it can respond to many of the needs of the inner areas and represent an opportunity for their sustainable regeneration (Melis, 2017).

A sustainable tourism plan that capitalizes and promotes local resources while respecting territorial fragility and uniqueness could also direct resources more selectively toward innovative projects that help the construction of shared entrepreneurship and that focus on identity and quality productions, witnesses of local culture, and tourism attractions.

Case Study in the Aosta Valley Region

This case study concerns the Aosta Valley Region, in the Gran Paradies that cover six municipalities, namely Cogne, Rhêmes-Notre-Dame, Rhêmes--Saint-Georges, Saint-Nicolas, Valgrisenche, Valsavarenche. Thanks to the use of open-ended interviews with some stakeholders in the area, the following coverage highlights what the expectations and the main difficulties encountered in the area covered by onsite observations in the implementation of the national strategy. The common feature of the area strategies developed so far consists in testing new welfare models that can ensure better liveability for citizens and rebuild entire local systems. Health care, education, and mobility are indispensable drivers for setting up revaluation programmatic proposals from below, thanks also to the administrative-managerial role. The municipalities of the inner areas become a laboratory for testing new socioeconomic models capable of ensuring better livability for citizens and rebuilding the health and education systems to guarantee basic assistance levels.

As previously already noted, the SNAI strategy focuses on the inner areas and aims to address the demographic decline, supporting the demographic increase in these areas, and to improve quality of life for the population of these areas, increasing well-being and offering a satisfying number of services (Barca et al., 2014). The procedure for inner areas requires the regions to study their own territory and, through discussion tables, to define the area of greatest need. Through this procedure, the Valle d'Aosta region decided to select the "Grand Paradis" area for its first case.

The area is based in Valle d'Aosta region, which is made up of six municipalities: Cogne, Rhêmes-Notre-Dame, Rhêmes-Saint-Georges, Saint-Nicolas, Valgrisenche, Valsavarenche.

The six municipalities of Valle d'Aosta region are located at high altitudes or the bottom of the valley and for these reasons they have different problems related to road access and the connections with the main town. The distance between Aosta and other municipalities ranges from 5 to 31 kilometers (Dipartimento per lo Sviluppo e la Coesione Economica, 2016). According to Preliminare di Startegia, the average population density of the area is low, around four inhabitants per km^2, and each municipality has experienced a demographic decline during recent years.

To realize the Strategia d'Area, the municipalities of the project commissioned different institutions to manage the strategy activities. The most important of these are the Unités des Communes Valdôtaines Grand Paradis, which represents a specific mountain area, the regional administration and the Gran Paradiso National Park, a protected area established by the state to preserve the ecosystems (of international and national importance) of the valleys around the Gran Paradiso massif for present and future generations. These institutions were also in charge of creating the "Preliminare di strategia," an essential document that contains descriptions of the territory and its peculiarities, the leading idea of the area, measures, and expected results from the SNAI.

The leading idea is to contribute to the development of a territory by supporting its ability to preserve and produce cultural and social value through the diversification of the tourist offer, agriculture, and crafts based on local authenticity, investment in culture and human resources, and international promotion of natural heritage (Dipartimento per lo Sviluppo e la Coesione Economica, 2013). These areas, despite their touristic and economic potential, are rural characterized by a low development and by an economic gap which can be reduced through measures, involved in the SNAI strategy, to support local development. Some of the expected results, outlined in the "Preliminare di Strategia" (2016), concerned the interruption of depopulation, the digitalization of administration processes, and the increase of the touristic movement through the seasonal adjustment and the growing of average stay. Each one of these goals established different actions to reach the targets.

After presenting the territory and its institutions and defining the leading idea, actions, expected results, indicators, desired trend, and timing through the "Preliminare di Strategia," the region defined and signed the final called "Accordo di Programma Quadro" in 2019. To investigate the realization of the SNAI Strategy and understand the main outcomes and impacts of the project on the territories, the main research questions concerned the satisfaction and the expectations of the area in relation to the strategy, the results, when achieved or not, and the management of the expected funds.

To answer these questions, three qualitative in-depth interviews supported the study and revealed the level of satisfaction about the strategy from the perspective of different local actors involved in the SNAI. Thanks to advice from a hotel manager based in Rhêmes-Notre-Dame, it was possible to identify and contact some of the main stakeholders involved in the project area related to the project. The President of the Gran Paradiso National Park, an institution that has a symbolic function for the project area, declared that the goals of the strategy were coherent and integrated with the socioeconomic development plan of the park. Also, this informant underlined how local institutions were poorly involved in the policymaking process.

Despite the limited intervention of the park, its President judged the strategy positively, as the municipality had the opportunity to integrate some financial resources to use in the achievement of shared goals. He was also pleased because finally the National Park was gaining a symbolic role and seen as an added value to the area development. Nevertheless, until today the President does not perceive any action, or updates about the evolution of the strategy in the entire area, except for some improvements in the education sector.

The second interviewee was a referent of the project area "Grand Paradis," President of Unités des Communes Valdôtaines Grand Paradis and Mayor of Arvier, one of the municipalities in the strategy area. He pointed out that the strategy was seen positively and with enthusiasm by the whole area from the very beginning. The municipalities, guided by the referent of the area, participated in the process, and outlined the main interventions needed in their territories. However: "interventions and expected results are clearly defined by the rules of European funds and this didn't allow to help in our territories craft, production and agricultural activities as we first imagined" (Second interviewee, May 8th, 2020).

Furthermore, the President also outlined some difficulties in practical and methodological issues like relating to teamwork and in planning long-term activities. Regarding the expected results and planned interventions, the interviewee explained that it was too soon to evaluate the strategy results. The reason is that the APQ (Accordo di programma quadro Regione Valle d'Aosta "Area interna – Grand–Paradis") Dipartimento per lo Sviluppo e la Coesione Economica (2016) was signed at the end of 2019, and the implemented actions were few.

Due to the stoppage caused by the current pandemic emergency, and the forced quarantine, it has been even more difficult to carry on the implementation of the strategy in this area. Currently, the intention, according to the second interviewee, is to carry out some of the measures pre-established before the end of the year in the health and transport sectors. In the President's opinion, it will be not possible to realize some of the actions by the end of the strategy implementation period. Regarding its pre-established funds, the second interviewee declared that until then the area had not received any financial resources. The few actions carried out became possible, thanks to some resources provided by the area itself.

The third interviewee was the Mayor of Cogne, one of the municipalities included in the project area. Cogne, among all municipalities, is the most touristic and populated and maybe, as the Mayor said, "the most peculiar" among the others involved in the project area. Cogne has different needs because it has a large population and is closer to the regional capital, Aosta. Considering the three essential factors which the strategy aims to reinforce, it is clear that Cogne needs divergent actions. Regarding the school sector, some municipalities included in the inner area have multi-classes, and Cogne itself does not have an education until middle school. Concerning the transport sector, Cogne has a connection with public transport every hour in every season, while the smaller municipalities have just one or two rides per day. Lastly, for what concerns the healthcare sector managed by the region, Cogne is the only municipality that has 118 volunteers on site (the third interviewee, August 31st, 2020).

During the interview, the Cogne Mayor confirmed the previous affirmations of the two other interviewees. In his opinion too, the strategy implementation is overdue and with few actions. The area has been a part of the strategy since 2015, when the team, composed of local authorities and government representatives, started to plan, and define premises and results, but the APQ has been signed only at the end of 2019. This leads to a critical issue concerning the bureaucratic procedure. According to the respondents, the procedures were too complicated and long, as it took four years to define the APQ and to start the implementation of the planned actions. In Cogne, the few actions activated consisted of training courses for teachers. As the Mayor repeated more times, the municipality did not perceive satisfying results until now. Furthermore, he added, "the municipality is still waiting for an electric vehicle that they intend to use as ski bus, shuttle or school bus."

Three conclusions are identified concerning the implementation of the strategy in this project area. First, the complex administrative and bureaucratic procedure led to a delay in the approval of the APQ and, therefore, to a delay in the planned actions. Second, the funds were limited to specific measures, not sufficient to satisfy the needs of the different municipalities. Third, the pre-established resources have never been distributed to the inner area and the actions implemented were possible due to the resources offered by the region itself.

These three critical aspects related to the strategy have been highlighted by the stakeholders and led to general dissatisfaction with the implementation of the strategy in this area. "The strategy had an external direction to the territory for which municipalities could not influence the decisions made. In fact, the administration for inner areas situated in Rome had a lot of influence" (The third interviewee, August 31st, 2020). This for the Mayor was an error because, even if they were experts, they did not understand the territory and his needs. Finally, another issue concerning the SNAI strategy is that it was evident that the project priorities were related to costs and financial resources and not to the real needs of the area (the third interviewee, August 31st, 2020). The implementation period is almost over and despite the various premises of the strategy, only a few have been reached.

With these first clues and reflections, it is possible to conclude that, even if the strategy was positively appreciated by the territory, it had several problems which led to the discontent that could be the cause of a lack of trust in institutions and proposed strategies. The development axes adopted by the National Inner Areas Strategy in Italy give sustainable tourism significant importance as a tool for activating local development paths. The national strategy for sustainable tourism also analyzes the potential of the inner areas as an implementer and active relaunch laboratory of the territories.

CONCLUSION

The process analysis of this chapter aimed to detect the main strengths and weaknesses in the implementation of a public policy to the Italian Inner Areas. For the moment, the implementation of actions on a territorial basis still lacks solid planning, a strong partnership between the representatives of the national strategy and local administrative actors and effective management of the funds allocated to the projects. The case of the Gran Paradis area illustrates some of these limitations, although the study recognizes that it is still too early to evaluate the effects of the strategy (Elliott & Salamon, 2002).

Another consideration regards the declarations of intent of the planning documents and area strategies, promoting the implementation of sustainable tourism actions, which contain risks and opportunities. The risks consist in supporting and intensifying productive and economic activities that neglect the real social and cultural vocation of the selected areas. In fact, while considering the importance of economic sustainability in local development projects, it is significant to connect this priority to social and environmental sustainability in coherent planning of the three dimensions.

Analyzing the approved case studies and the general strategy, it is also important to underline the importance of the construction phase of the connections between the central and peripheral areas. The construction of these connections can generate positive externalities for the inner areas deriving from the construction of integrated tourism activities with the poles. In this way, attractive poles equipped with greater economic and productive tools could facilitate tourism in the inner areas through routes that lead from the center to those areas rich in social and cultural capital.

The opportunities consist of the intangible resources present in the area that should be enhanced and promoted by the Inner Areas Strategy means through training activities for the on-site staff. Tourism agents and local development representatives would thus be provided with sustainable tools and methodologies capable of grasping the complexity of the territory to structure an effective action strategy. Planning and renovation projects of inner areas should benefit teams equipped with sociological and geographical skills that allow a systemic reading of the territory, its peculiarities, its social, relational, cultural, and natural capital. In these contexts, tourism cannot act as a mere economic activity since it is an agent of great cultural impact capable of redesigning and altering the state of the places.

The policymaking process of the National Inner Areas Strategy must capitalize on these territorial potentials and train a class of operators capable of carrying out sustainable development projects for these areas, which, although apparently disadvantaged, can prove to be rich custodians of tangible and intangible resources. In the light of these last considerations and returning to the initial question of whether or not sustainable tourism is an oxymoron, it is possible to say that sustainable tourism can very often turn out to be a paradox in practice: the actors, belonging to different territorial levels, if excessively conditioned by economic and political interests, are unable to plan and implement tourism strategies embracing the principles of sustainability. On the other hand, however, looking at the case study, the people interviewed manifest a sincere interest in achieving the main aims expressed in the national strategy and there was often a desire to overcome political stalemates.

The issue, therefore, is once again a cultural problem and the effort to make sustainable tourism a reality and not an oxymoron must begin with training tourism professionals, especially as facilitators of cultural change. These professionals must embrace the principles of sustainability and "fight" to implement them in practice. In other words, it is important to make these professionals resilient. They must be capable of not letting themselves be overwhelmed by discouragement when (inevitable) conflicts arise, to be focused on their objectives, and to be aware that the change is possible and above all inevitable to face the climatic challenges, which arise continuously and daily, and which the tourism sector must also take on board.

Chapter 3

FROM FORDISM TO TOURISM Genoa, Turin, and Milan

Monica Bernardi
University of Milano-Bicocca, Italy

Ezio Marra
University of Milano-Bicocca, Italy

Abstract: This chapter examines three Italian cities that have experienced a transition from "Fordism to tourism": Genoa, Turin, and Milan. After an industrial crisis, they have invested in culture and tourism as alternative ways of development. This transition is examined using the theoretical framework of urban regimes highlighting five development trends: the city as a growth machine, the Fordist city, the creative city, the city as entertainment machine, and the blue-green city. By adopting this theoretical framework, the evidence shows how academic institutions, tour operators, and public authorities may or may not work together for the tourism development of their cities. **Keywords**: Urban regimes; post-Fordist tourism; Milan; Turin; Genoa; city-tourism

INTRODUCTION

This chapter aims to understand the transformation of Genoa, Turin, and Milan from industrial conglomerates to tourism cities, relying on the theoretical framework of urban regimes to understand their transition from Fordism to tourism. The five regimes presented are not considered to be successive stages of urban development; they can coexist and compete. The three cities were chosen since they are part of the Italian industrial triangle and have been the engine of the country. Toward the end of the last century, the cities experienced a crisis of their industrial model, initially perceived as

Managing Destinations: From Theories to Practices
Bridging Tourism Theory and Practice, Volume 14, 23–39

Published under exclusive licence by Emerald Publishing Limited
ISSN: 2042-1443/doi:10.1108/S2042-144320240000014003

temporary (Bagnasco, 1977). When these cities realized that the issue was structural, they tried to relaunch their productive capacity. The cities converted their productive structure toward new information and communication technologies as well as culture and art. Tourism, initially, was not perceived as a source of new economic, employment, and cultural opportunities. In the early 1990s, cities began to grasp the challenges posed by economic globalization and the shift to a post-Fordist era. In order to compete with other major European and global metropolises, they adopted a new urban agenda that recognized tourism as a driving force for development (Marra, 1989).

Genoa was the first city to realize its industrial crisis. It tried to reposition its development by investing in the *Colombiadi*, the celebration for the 500th anniversary of the America's discovery (1992). Then Turin, a well-known Fordist company town focused on an automotive monoculture, attempted to reposition its image by competing for the 2006 Winter Olympics. In Milan, the industrial restructuring process dates from the 1980s. Since then, it has become the capital of fashion and design, investing mainly in business tourism, but also EXPO 2015 has changed this orientation and the external image of the city (Ferrari & Guala, 2017).

In reading the transformation of these cities into tourism destinations, two key facts of Italian history should be considered as they have changed the way urban governance is perceived. The first is the direct election of the mayor by citizens in 1993 (law n. 8 of March 25, 1993), previously elected by the city council. The second is the so-called "Delrio law" of 2014 (n. 56 of April 7, 2014), which reformed local authorities, introduced the metropolitan cities, and redefined the provinces. The operation and effectiveness of metropolitan cities is a work in progress and the slowness of change has affected the tourism development policies of these three cities.

According to the National Institute of Statistics (ISTAT, 2020), tourism in Milan grew during the pre-COVID-19 era, overtaking Venice and Florence, to become the second Italian city for overnight stays after Rome. Turin and Milan recorded their highest level of tourism development between 2008 and 2017, while Genoa had the lowest in that decade. At present, the pandemic has imposed travel restrictions that have drastically reduced tourist flows, allowing a reflection on what may be most appropriate regime.

Urban Regimes

The theory of urban regimes was developed during the 1980s. It suggests that in fragmented urban contexts, government cannot be conceived as a form of hierarchical control by elected decision-makers, but mainly as an arena of cooperation between different actors (Ostrower & Stone, 2006; Stoker, 1995; Stone, 1993). An urban regime is a governance structure able to include public, private, and social actors in the defining of strategic agendas (John & Cole, 2000).

After the crisis of the Fordist and Keynesian models due to the impact of globalization, especially within industrial cities, urban decision-makers began to look for new opportunities to avoid the economic and social crisis by converting their development model (Belligni & Ravazzi, 2013). In a weak political context with a crisis of financial resources, local political élites sought new alliances with relevant social actors to establish hegemonic and successful coalitions to produce economic or social development, the former growth-oriented and the latter welfare-oriented. However, coalitions that generate urban regimes should take into account five different types of development models, each involving different actors and stakeholders pushing in different, often conflicting, directions (Davies, 2017). Five urban regimes, each with its sub-agendas have been identified (Belligni & Ravazzi, 2013).

The City as a Growth Machine. This concept was defined by Molotch (1976) and considers the role of landowners, élites, entrepreneurs, and land-related interests (big banks, real estate agents, lawyers, public services, department stores, and metropolitan newspapers) in promoting, through local government, the growth of the city (Cox, 2017). Molotch and Logan (1987) argued that the commodification of places and the political economy influence the growth of cities. This may be achieved through land monopolies and collective action, where entrepreneurs secure control of property for profit based on its exchange value. The linking of élites with local and national government fosters pro-business policies that ensure the success of the city's economy. As Anselmi and Vicari (2019) point out, party bureaucrats still play a critical role in shaping the city and in negotiations between political parties and private interests. Within this scheme, there are two sub-agendas: real estate and land use.

The Fordist City. This concept refers to the standardized and industrialized mass production system promoted by Henry Ford to meet the demands of a mass market (Farkas & England, 1988; Gramsci, 1992). Within cities, there has been a resurgence of manufacturing, albeit in digital form, with the aim of bringing production back. From the dismantling of the city-factory, there is a movement toward urban manufacturing that increases the appeal of cities. According to Sassen (2008), the new urban manufacturing reverses the historical relationship between services and production where services have historically been developed to meet the needs of manufacturing companies, while in this case the industry produces on demand for the service sector.

The Creative and Polytechnic City. This regime (Bagnasco, 1990; Florida, 2002; Landry, 2000) is linked to "cultural," "creative," or "knowledge" economy and, more recently, to the diffusion of information and communication technologies and the emergence of the smart city (Bria & Morozov, 2018; Hollands, 2008). Landry (2000) advocated a need to develop a culture of creativity within the bureaucratic structures that govern cities. This would include a review of regulations and incentives that support creativity as an asset for wealth creation, and to legitimize the use of imagination within the public sphere. Florida (2002) promoted the creative class based on "talent, technology, tolerance" and linked this to a "creative economy" that favors the development of a city driven by lifestyle factors, urban infrastructure, and entertainment.

The City as a Leisure and Entertainment Machine. This regime (Lloyd & Clark, 2003; Marra, 1989; Zukin, 1982) has three sub-agendas: art and culture (Zukin, 1995), mega-events (Guala, 2015), and tourism (Shoval, 2018; Spirou, 2010). The focus is on the capacity for consumption, culture, and entertainment to change cities. In Clark's view, services and culture guide urban policies, rather than the opposite: "Consumption ... is becoming a critical focus of politics, nationally and in cities, complementing more traditional issues of production, work, and even taxes" (2001, p. 4). This position is partially embraced by Zukin (1998), who showed how three notions of culture (ethnicity, esthetics, and marketing tools) are reshaping urban places and conflicts of revitalization. Mega events and tourism are also part of this scheme as the promotion of leisure time influences urban development.

The Blue, Green, and *Blue-Green City*. This regime (Owen, 2009; Rozos, Makropoulos, & Maksimović, 2013; Wrenn, Casazza, & Smart, 1983) is based on the promotion and creation of green areas, environmental sustainability, and water management (Davico, Mela, & Staricco, 2009). The focus is on the blue infrastructure, essential for public health and quality of life within cities and green infrastructure (parks, recreation areas, gardens, green roofs, and the like) that provides a range of urban ecosystem services. These systems can also be useful for flood, noise, and drought mitigation and reduce air pollution and urban heat effects. The integration of the two assets promotes local development, drives urban revitalization, and attracts tourist flows.

A winning urban agenda is derived from an alliance between two or more different regimes. The emerging coalitions are, therefore, made up of different mixes (which vary from city to city among the different models listed). The coalitions that generate a winning urban regime are often the bearers of legitimate (and sometimes illegitimate) interests that prevail over losing coalitions. The combination of the different models generates agendas and sub-agendas between the cities examined. Tourism is particularly linked to the city as a leisure and entertainment machine model, but cannot be separated from the blue, green, and blue-green city, especially in the current era and in consideration of the Sustainable Development Goals.

Methodology

A mixed method approach allowed the identification of each city stakeholder coalitions within each regime, the current trends and the main interventions on the political agenda. This approach guarantees greater flexibility and richer and more complete data, which limit the contradictions between quantitative and qualitative results (Creswell & Plano Clark, 2007). Secondary data were also collected through Instituto Nazionale di Statistica databases and official sources in metropolitan cities. These include official documents of metropolitan cities with their related datasets, maps, strategic plans, and reports (Giorgio Rota Reports, the Ispra-Snpa Reports, the IRES-ARPA Report, as well as the Sustainability Reports for Turin, the 2019 data by the Osservatorio Assolombarda Milano and Centro Studi, and the 2019 Rapporto Ambrosianeum for Milan, Chamber of Commerce Reports for Milan, Turin, and Genoa). The analysis included the electoral programs of the current city mayors, and a set of short interviews (in the form of colloquium) with some of the main experts and stakeholder representatives (10 for each city).

THE ITALIAN INDUSTRIAL TRIANGLE

For a long time, the northwest regions of Italy polarized around Genoa, Turin, and Milan, shaping the so-called "industrial triangle" or the industrial engine of the country, expression of the national economic potential, and driver of the sociopolitical orientations (Berta, 2008). It represented the epicenter of the "economic miracle" (Bagnasco, 1986), the "hard core" of industrial development, connected with the "working class." According to the 1931 census, every 100 people employed included 47 workers in Genoa, 46 in Turin and 43 in Milan, respectively (Arvati, 2005). But the real celebration of the industrial triangle came with the economic boom between the 1950s and 1960s. The book *Tre Italie* (Three Italys) by Arnaldo Bagnasco in 1977 compared the center and northeast Italy based on widespread industrialization, with the radically different northwest Italy, based on the dominance of large companies. Bagnasco's work was a celebration of the "First Italy," coinciding with the industrial triangle, and in his view, the main characteristics of the three northwestern cities are: large companies, high-intensity capital, greater use of technology, and the strategic nature of production compared to other sectors (Bagnasco, 1977).

Despite their common characteristics, the three cities had different developments, and in the 1970s, they appeared very different from each other. According to Bagnasco,

> Turin is the unique city in the sector that drives post-war development ... Milan is already a more differentiated and complex situation, which finds its central role also and above all at the most abstract level of power and financial intermediation. Genoa is a commercial capital and a large state industry, which is seeing its production structure seriously eroded over time, and which is losing its financial and executive functions. (Bagnasco, 1977, p. 220)

According to ISTAT (1973, 2003, 2013), as seen in Table 1 below, between 1971 and 2011, there was an emblematic change of pace with the loss, in Italy, of about 1.2 million jobs in the manufacturing sector, 900,000 of them lost in the industrial triangle: in absolute terms Turin, Milan, and Genoa lost 282,000, 563,000, and 59,000 jobs, respectively; the three provinces of the triangle alone accounted for almost 27% of all manufacturing employment in Italy in 1971, around 19% in 2001, and less than 12% in 2011. In a relative sense, Turin in 2001 accounted for 54.5% of manufacturing employment, then drastically reduced to 35% in 2011; Milan went from 48.3% in 2001 to almost 20% in 2011; and Genoa from 48.5% in 2001 to 24% in 2011.

However, this reduction in manufacturing jobs has been largely offset by strong job growth in the services sector because of the so-called *tertiarization* that has marked mature industrial companies in the deindustrialization phase. As pointed out by Krugman, all industrial countries have experienced this phenomenon as a combined effect of the rapid increase in productivity and the decline in demand, which has led to a decrease in industrial production and employment. However, the effects due to industrial reorganization must be distinguished from those due to the current economic crisis, which affects all sectors. Consumers have gone from buying tangible goods to buying typically intangible services (including those for tourism and leisure), reducing demand, which combined with the productivity of the sector, has reduced the number of workers needed to meet the demand for goods (1994, pp. 263–264).

The effects of deindustrialization are evident in these three cities, developing economic repositioning strategies that have transformed their image at national and international level. All three have tried to diversify their main economic and productive activities toward the tertiary sector, but also promote local marketing to attract investors and resources. The renewal of their urban layout became a prominent feature, touching not only the communication dimension and the service system but also urban decor and the culture/events industry. These processes of urban regeneration also changed the position of cities in the general market: their interlocutors were no longer only economic actors interested in long-term investments but also those "urban populations" (Martinotti, 1993) who want to lead their city life in different ways and with a different pace: students, tourists, day-trippers, and consumers of various kinds. In this sense, the end of the industrial phase let cities realize the importance of including the tourism dimension as an incentive to become more attractive (Marra, Mela, & Zajczyk, 2004). In this complex transformation, from Fordist to tourist cities, all three have invested in distinctive events to attract capital, interest, and public attention internationally.

Genoa was the first in 1992 with the *Colombiadi*, opening a new season of what Amendola (2003) called the "postmodern city of the new millennium": a city where esthetic ethics were increasingly affirmed due to famous architects, who regenerates the city to attract economic actors and investments.

Table 1. Manufacturing Workers 1971–2001, Provincial Data.

	Workers 1971	Workers 2001	Workers 2011	Workers 2001 (1971 Base 100)	Change in Workers 2001–1971	Workers 2011 (1971 Base 100)	Change in Workers 2011–1971
Turin	476	259	194	54.4	−239	34.8	−282
Milan	789	381	226	48.3	−598	19.5	−563
Genoa	97	47	38	48.5	−111	23.9	−59
To-Mi-Ge	1362	925	458	67.9	−948	24.5	−904
Italy	5102	4853	3892	95.1	−293	75.0	−1210
% manufacturing workers in Italy	*26.7*	*19.1*	*11.8*				

Source: Elaboration from Istat Data (x 1000).

Turin was host to the 2006 Winter Olympics (Guala, 2007) and Milan held Expo 2015 (and will host the 2026 Winter Olympics).

From their relaunch on the international tourism panorama, the three cities have gained popularity and attracted growing tourist numbers, reaching record peaks in 2019. The Milan Municipality alone received close to 7.5 million tourists, 9.4% more than in 2018 when 6.8 million people spent at least one night in the city (the metropolitan area reached almost 11 million tourists) (Milan Municipality, 2020). Turin has touched almost 1.5 million tourists, recording an increase of 6.6% compared to the previous year (data from the 2020 annual report of Torino Turismo official). While in Genoa, according to the Regional Tourism Observatory (2020), the number of tourists increased by 15.7% compared to 2018 (despite the collapse of the Morandi Bridge).

With the global crisis, the need for "less esthetics and more ethics" increased, pushing cities to open more space for integrated and sustainable bottom-up socioeconomic growth policies. With COVID-19 resetting tourism, the relaunch allows reflection on how a city may still be attractive and in which way. The three cities have signed a memorandum of understanding to intensify their institutional relations through participation in projects related to economic and tourism development, cultural activities, and joint tourism promotion.

Starting from the information and data collected through institutional and official reports and thanks to the informal colloquiums, each urban regime is here more closely investigated for each city.

The City as a Growth Machine

Genoa. In terms of large infrastructure, the metropolitan area of Genoa (and the entire Liguria) suffers a lack of connection opening to broader issues at the national level. These include the relaunch of the port of Genoa and of the airport, to serve Genoa and the greater Genoa and Riviere region. This aimed to ease metropolitan mobility and trigger an intense development process. A negative aspect of the metropolitan area is *Rapallization*, a term used in the 1970s to describe the indiscriminate urbanization that occurred in many touristic Italian areas following the economic boom. Rapallo, in the metropolitan area of Genoa, symbolizes this building speculation. The collapse of the Morandi Bridge in 2018 also had negative economic, tourist, social, and emotional consequences (Rania, Coppola, Martorana, & Migliorini, 2019). However, the city reacted quickly with a reduction of less than 1% in the industrial sector, the employment rate has increased by 0.6% (Lingua, 2019), and the new bridge was inaugurated in August 2020, opening new opportunities for the metropolitan area. The bridge reconstruction involved a famous local *archistar*, the Pritzker Prize winner Renzo Piano.

Turin. The urbanistic plan elaborated by Gregotti and Cagnardi includes several programs for major urban works, an emblematic one is the skyscraper designed by *archistar* Fuksas that will host the Piedmont Region Palace by the end of the 2021. The new Porta Susa Station portion of the city has been renewed inside the major urban redevelopment project of the Spina Centrale (Saccomani, 2019). Other major architectural and urban interventions include the new Officine Grandi Riparazioni with 90 million euros invested in the restoration and revival of the historic railway vehicle factory; it is the largest philanthropy venture project in Europe, and oriented toward growth and economic, cultural, and innovation development of the area (Amione, 2018).

Another major intervention is the Cittadella Politecnica of Design and Sustainable Mobility which stems from the need to redevelop structures and training programs within a single building complex integrated into the city. The university has also planned a Citadel of Space and a museum technology center developed by Altec and Thales Alenia. The third intervention is the Intesa San Paolo skyscraper, "Building of the Year 2016." It is a new green and sustainable tower designed by Renzo Piano based 100% on renewable sources.

The Lingotto, designed by engineer Trucco for FIAT, was a key example of Italian modern architecture, and in 1983, was redeveloped by Renzo Piano. This redevelopment added a complex system of functions and integrated public spaces (the Auditorium, an art gallery, a multiscreen cinema, and a large shopping mall). The Lingotto is still undergoing redevelopment with the opening in 2007 of the first EATALY shop (a slow-food and high-quality food presidium planned by Oscar Farinetti which also opened in Genoa in 2011 and in Milan in 2014), the transformation of the 8Gallery shopping center, and the greening of the runway, a Turin high line. The renovation of railway stations (started in 2019) is also interesting with seven former railway areas (for a long time in a state of decay) being transformed into spaces for tourism, accommodation, trade, and the service sector. In terms of land use, according to the 2019 Ispra-Snpa Report, while all Italian cities continue inexorably to consume land, Turin reversed the trend and began to recover land (seven hectares of land were regained in 2018).

Milan. Costa, Cucca, and Torri (2016) described Milan as a city "lost in the transition" from the paradigm of the growth machine to a social innovation approach. The city has been governed for decades by a strongly market-oriented system of governance, and now by an administration oriented toward social innovation in response to the growing inequality (Dente, Bobbio, & Spada, 2005). However, infrastructure projects still dominate, for example, City Life, a redevelopment project in the Milan Fair district, characterized by prestigious buildings, new luxury apartments, and the largest urban commercial district in Italy (Bonfantini, 2019). The complex was designed by the three-time Pritzker prize-winning *archistars* (Zaha Hadid, Daniel Libeskind, and Arata Isozaki).

The regeneration around Porta Garibaldi station was designed by *archistar*, Gae-Aulenti and the Porta Nuova project by Cesar Pelli, Stefano Boeri, and Nicholas Grimshaw (including Boeri's vertical forest). The program for the rehabilitation of the former railway stations was also recently approved and seven disused railway stations will be redeveloped (Farini, Porta Romana, Porta Genova, Greco-Breda, Lambrate, Rogoredo, San Cristoforo). This will become the largest urban redevelopment plan of the city in the last 20 years. Noteworthy is the Rem Koolhaas' project (another Pritzker prize), Fondazione Prada's project and the new Fondazione Feltrinelli by Herzog and de Meuron (Pritzker prize). Looking outside the city center, Milan also seeks to enhance its relationship with the hinterland, strengthening transport infrastructures and upgrading the services and image of all minor centers, to create urban continuity.

This growth machine realigns and interacts with the other urban regimes, creating buildings for culture and leisure that become icons of possible tourist attraction.

The Fordist City

Genoa. This city is a port and industrial center. Since the late 1970s, it experienced a drastic reduction in port activities and then the dismantling of state-run industry because of new neoliberal policies (Ferrari, 2007). Bertelli (1991) reports that at the provincial level, between 1981 and 1990, Genoa recorded a loss of a quarter (24%) of jobs in the state-run industry. Port activities were in crisis due to both economic stagnation and restructuring-related conflicts. In the mid-1980s, the old port area, the ancient dock, and the 19th century warehouses were completely abandoned. However, in the 1990s, the port became competitive again by adopting container loading/unloading equipment, rapidly increasing goods handled with containers from 5% in 1980 to over 20% in 1997.

According to the 2017 Metropolitan Strategic Plan, manufacturing contributed 10% to the total added value in Genoa, against 12.4% in the other cities in the center-north and 16.5% for the entire country. Genoa's abdication of its industrial vocation can also be found in the dynamics of added value in the period 2001–2011. Industrial activity in the metropolitan area grew by 5.2%, far from the percentage change recorded in other metropolitan cities and in those of the center-north. During this time, the manufacturing sector of metropolitan Genoa declined by about 5%, while in the remaining metropolitan cities, on average, there was a positive change.

Turin represents the Italian working-class city par excellence. It is a Fordist city characterized by a social reality closely linked to a manufacturing entirely focused on the production of cars. This connotation has been reinforced and celebrated by the cultural tradition of the workers' movement (Vanolo, 2015). For this reason, the transition "from Fordism to tourism" here is more emblematic than in Genoa and Milan, even if the changes were mostly the same. The first symptoms of deindustrialization date back to 1971 and have seen the industrial sector halved. For a long time, the phenomenon was linked more to the outsourcing of industries than to the emergence of new production areas. However, since the 1990s, job creation in services and tourism has been dissociated, at least in part, from industrial demand. Today, the city is experiencing a manufacturing renaissance, with the 2016 report "From the dismantling of the city-factory to the new urban manufacturing," published by Censis (2016), demonstrating that Turin was the most "manufacturing" city in Italy, with the highest concentration of activities in the urban area. The Lingotto is seen as a temple of Fordism and modernity.

Milan. At the end of the 1970s, Milan was very different from the other two, with a much more diverse and mixed social and political composition. It has been characterized by being a bourgeoisie made up of a mix of corporate profit and private income with overlaps between the public and private spheres. It has a high level of economic self-sufficiency, internationalization, and has links with financial capital. According to Marra (1989), to access business lending, companies largely depend on public banks often managed by local and national political powerbrokers. At the end of the 1970s, fractions of the old bourgeoisie and parts of the middle class represented a block of "property owners," with great lobbying power tending to replace business profit with property income, sometimes on a large scale.

The process of deindustrialization led to the transformation of Milan into a managerial city, international capital of fashion and design, headquarters of several trade fairs, in competition with cities such as Frankfurt in the exhibition panorama. The industrial history traced by the big bosses such as Falk or Pirelli was kept alive converting industrial areas into business incubators and locations for small firms, as well as for cognitive and research centers, as in the case of the Milan-Bicocca University, located in the former site of Pirelli (dell'Agnese & Anzoise, 2011) and planned by the *archistar* Gregotti. Today, one is witnessing a revival of manufacturing based on technology with the idea of bringing production back to the city: emblematic is the project "Manifattura Milano" which promotes the development of new craftsmanship, digital production, and Industry 4.0 (D'Elia, 2018).

The Creative City

Genoa. It took many years for Genoa "to accept the idea that the era of heavy industry was over, but in the end the city did and began to shift its hope of an economic recovery to the ICT industry that had spontaneously created a hi-tech cluster in Sestri Ponente (one of the oldest industrial districts in Italy)" (Bobbio, 2005, p. 6). The city has been an active pioneer in the smart city, technological innovation, and energy and sustainable mobility policies since the foundation of the smart city sector in 2012. It then evolved into the Smart City Strategies and Supranational Projects office. The role of the Genoa Smart City Association was crucial, at the municipality and metropolitan level, and inspired the Scope Document on urban adaptation to climate change proposed by the city and approved by the European Commission. The five pillars of the Genoa smart city strategy are a digital agenda, sustainability of public services and urban mobility, services for social inclusion, infrastructures for social and technical assistance (Genoa Metropolitan City, 2017).

Turin. The Officine Grandi Riparazioni of Turin, discussed earlier, once fully operational, will host exhibitions, shows, concerts, theater, dance and performing arts events, workshops, start-ups, innovative companies, combining the ideas and values of creativity with the tools and languages of new digital technologies. A center for visual and performing arts and scientific, technological, and industrial research, the Officine Grandi Riparazioni will act as a sounding board for the rich fabric of the cultural and productive realities of the territory (Martini, 2011). The smart city approach was based on a master plan called "Smile" proposed by the City of Turin and the Turin Smart City Foundation (SMAU, 2014). Smile stands for smart mobility, inclusion, life and health, and energy. It outlines a path for the transformation of the city including efficient and nonpolluting mobility, the rational use of energy from renewable sources, a society open to people's needs and health balanced with the help of new technologies to support a higher quality of life and attractiveness for tourists and investments and an increasingly efficient public administration. In addition, the city is extending its Wi-Fi system and launching the use of 5G.

Milan. According to Assolombarda (2016), Milan has three main trajectories of innovation: life sciences (biotechnology, biomedical, and pharmaceutical technologies through a network of hospitals and research and care institutions are a key resource for the development of the city); production 4.0; and art, culture, and creativity.

In 2013, the municipality launched a participatory study and reflection process to define the administration's guidelines on the smart city, which ended with the "Milano Smart City" document approved by the City Council in 2014 (Milan Municipality). This document highlights the importance of combining economic development and social inclusion, innovation, training, research, and participation. It is organized as seven laboratories on the topics of the global, national, and European city; sustainable urban mobility; environmental and energy policy; social inclusion and diversity; city welfare; simplification of public administration; and business generation. Within the smart city umbrella, the municipality has also developed the "Sharing City" project, and it is active in smart and sharing oriented projects such as shared urban gardens, maker-spaces and Fablabs, coworking incentives, civic crowdfunding initiatives, incubation programs, and the like.

The Metropolitan city is also developing several smart projects, such as the "South-East Smart Land" project which involves 14 municipalities willing to support the evolution and development of the territory through innovation, sustainability, cohesion, and contamination. Other projects are the Esmartcity (Enabling Smarter City in the MED Area through Networking), focused on the creation of innovative ecosystems by citizens, businesses, universities, public bodies, and existing clusters; "+Community" for the simplification of public administration; "Smart Edge" for the energy efficiency of buildings; "Digital Campus," "5G," and "Nautilus" related to connectivity.

The City as a Leisure and an Entertainment Machine

Genoa. The first events outside the industrial and port perspective were the 1990 World Cup and the 1992 *Colombiadi* (the celebrations for the 500th anniversary of America's discovery). These celebrations supported the redevelopment of the old port area and part of the historic center. It was also an opportunity to expand the underground (opened for the 1990 World Cup). In later years, Genoa hosted the 2001 G8 Summit, and then, in 2004, was nominated the European Capital of Culture. This last event represented an opportunity for the city to show itself as a city of art and culture and to highlight its historical heritage (Storti, 2015). The program of temporary, blockbuster exhibitions was aimed at attracting tourists (Sacco & Blessi, 2005). The Old Port area is now one of the main tourist attractions, and includes the City for Children, the Aquarium, hotels, shopping areas, and leisure activities. The city center is also doing well (Lourenço & Bardi, 2005). Although industry still employs 24% of all employees, the city is increasingly evolving toward tourism and cultural activities. Transport links in Genoa, due to its peripheral location, suffers from poor accessibility, compromising its development. According to the Statistical Yearbook 2017 of the Municipality of Genoa (Genoa Municipality City, 2017), in 2016, the number of tourists to the Genoa Aquarium increased by 17.6% from the previous year. The Fair attracted 650,178 participants (0.9% growth compared to 2015). In 2016, city museums recorded 13.7% growth in tourists compared to 2015, the highest value reached. The Galata Museo del Mare was the most visited museum followed by the New Street Museums and the "G. Doria" Natural History Museum.

Turin. The nomination in 1998 to host the 2006 Winter Olympic Games allowed the city to become a new tourist and commercial hub displaying its history, culture, and high-tech industry. As reported by the Olympic.com website (2014), "in the two years after the Games ... the venues in Turin Olympic Park saw 187 events and 577,500 spectators use them for corporate, leisure, and sports events, while over 55,000 people attended entertainment activities." Turin has also become an important destination for corporate congresses, and the Olympic Villages built in the mountain resorts of Sestriere and Bardonecchia have been converted for tourist use. The urban repositioning of Turin and its metropolitan area included transport service (road and rail connections and the enlargement of the local airport), restoration of monuments, development of online tourism and cultural resources, and media visibility.

Thanks to the legacy of Turin's Winter Games venues, the city has also been able to attract a growing number of major events including the World Fencing Championships, the Winter Universiade (university games), the World Figure Skating Championships, the Turin World Book Capital, the 2008 World Design Capital, and Italia 150 in 2011. The city has also invested in permanent cultural industries such as the Book Festival, the Film Festival, the Slow Food fair, and the MITO Festival. This last event is evidence of growing cooperation activated between Turin and Milan. The festival originated in Turin at the end of the 1970s as "September Music" (a Baroque music festival), then it was extended to contemporary music with a robust international program in cooperation with Milan. This cooperation gave the festival a stronger identity within the European music festivals panorama (Bollo & Carnelli, 2008).

Milan. The city's creativity profile contributes to the entertainment machine and to the fashion and "Made in Italy" market. Milan is the main Italian capital of fashion and design, attracting tourists in these business sectors (Lodigiani, 2019). This has seen trade-offs in terms of seasonality (it has a higher performance during the week and especially on the weekdays of trade fair events, while hotel occupancy is lower on weekends and nonworking days). In 2015, Milan hosted a mega event, the Universal Exposition "Feeding the Planet, Energy for Life," which represented an important opportunity to reposition the destination in the leisure segment. Recent studies have confirmed the positive effects generated by the Expo on the performance of local companies and the ability to reduce the seasonality of the destination (Sainaghi & Baggio, 2019, 2021).

The Blue-Green City

Genoa. The 2017 Strategic Plan presented the new Port System (Legislative Decree 169/2016) based on the Western Liguria Sea Authority responsible for the Blueprint project. The project aims to renovate the port area between Calata Gadda and Punta Vagno redesigning a new promenade that includes the Pilot Towers, the industrial port, new industrial and urban spaces, and the waterfront within the port. Genoa was ranked n° 73 out of 104 cities in the 2019 Urban Ecosystem by Legambiente-Ambiente Italia and Il Sole24Ore.

Turin. The Metropolitan City Strategic Plan foresees the realization of a sustainable water cycle, the implementation of integrated urban waste management, and the adoption of sustainable mobility forms (Urban Sustainable Mobility Plan – PUMS and Cycle Mobility Plan – Biciplan foreseen interchange between private and public transport, the extension of line 1 and the building of line 2 of the underground) (Turin Municipality, 2013). The Passante Ferroviario represents an opportunity for urban regeneration to link the two parts of the city previously separated by the railway line. However, several reports highlight high pollution levels in the city. The 2019 Urban Ecosystem list of Legambiente-Ambiente Italia ranks Turin in 88th position (on 104 total cities). Lastly, there are very few initiatives concerning the Po River.

Milan. There has been much discussion about the creation of a green belt around the city, also called Orbital Park, Green River. The concept aims to connect Agricolo Park, Nord Park, Lambro Park, Forlanini, Trenno, and other green parks that together form a "green" orbit 30 times the size of Sempione Park, at 72 kilometers in circumference. While this vision is attractive, there have been no concrete steps undertaken.

Another debated project supported by the current mayor concerns the reopening of the Navigli canal. The project is on standby after a public debate, as there are other priorities for the city. Finally, Milan champions sustainable mobility, with various electric vehicles, bike, and car sharing solutions. The 2019 ranking of the Urban Ecosystem of Legambiente-Ambiente Italia placed Milan in 32nd position (out of a total of 104 cities), higher than Genoa and Turin. Noteworthy projects include the Boeri vertical forest (which won several international prizes) and the eco-sustainable Unicredit Tower, the tallest skyscraper in Milan and Italy.

As one can note, sometimes the cities' initiatives can fall in more than one regime and several overlaps emerge.

CONCLUSION

The data presented for each city allow us to trace some considerations on their tourism policies. The assumption at the start of this chapter that the five regimes were not incremental, or successive stages of city urban development, is partially contradicted. The assumption could be reformulated considering the regimes as partially coexisting and competing with each other in order to affirm themselves.

The analysis of the three cities shows how they have followed, even with different times, modes and outputs, the unfolding of the various regimes. Even if none of the cities have followed the regimes in a consequential way – but with continuous overlapping among the five models – the attempt of the three cities to relaunch themselves investing on different dimensions is evident; each time, each regime has tried to affirm itself and orient the city developments through its stakeholders and its interests to let the city grow and become more attractive; today, actors and initiatives related to the different regimes and agendas/sub-agendas are clearly mixed in the process of making the city attractive; all of them contribute to shape also the tourism attractiveness of the city, even if with different weights and modes. In this way, as a result, three cities that have never historically been tourism cities have managed to establish themselves in the national and international tourism scene.

Although there is no incremental and progressive sequence of the five regimes, at the end, they had inevitably converged on tourism issues to promote the local urban development and gain positions in the international and national tourism panorama, using creativity, leisure, entertainment, and all the necessary assets, thus integrating the five regimes. The regime of the "City as Leisure and Entertainment Machine" in some way has taken over, even if it has done so within a rapidly changing society. As the Spanish Minister of Industry, Trade and Tourism, María Reyes Maroto, recently said, European tourism "must ensure that it remains aligned with the changes that take place in our societies" (Luebke, 2020).

The need to converge more intensely on the last urban regime presented, that of the blue-green city, is increasingly clear. Cities are actively working to improve policies that promote green and blue infrastructures as an integral part of a local/regional natural heritage preservation strategy and to increase the presence of nature and water within their borders. The three cities testify to this new emerging approach and the importance of sustainable urbanism (Flint & Raco, 2012). It is also reflected at the tourist level, as tourism has the potential to contribute, directly or indirectly, to sustainability, to the 17 Sustainable Development Goals. Contemporary cities today cannot hypothesize their own development without acting in sustainable manner and this implies that tourism must be rethought in terms of sustainability. Genoa, Turin, and Milan are already working to meet contemporary tourists needs (more and more oriented to slow and green forms), and to face the Sustainable Development Goals.

As the COVID-19 pandemic has all bet halted tourism, it calls into question the leisure and entertainment regime, especially when disconnected from the green and blue dimension. The COVID-19 pandemic has demonstrated the importance and the need for a city to reorient its policies toward new issues, and to shape new perspectives of local development and attractive capacity. The three cities have experienced the annihilation of the tourism industry (Figure 1), where Milan in 2020 lost 61.6% of visitors, Turin 54.7%, and Genoa 60.4% compared to 2019 (National Tourism Agency, 2021).

Recently, the European Union has declared that the three points of action in the context of pandemic should be safety, sustainability, and data. It highlights the importance of ensuring a safe tourism experience, the promotion of greener vacations, and a move toward data-driven tourism. The Sorrento General States of Tourism, held in February 2021, stated that tourism should move toward a safe, green, and digital transition. This is important also to alleviate the effects of future crises and shows how in the future exponents of different regimes will be called upon to cooperate to

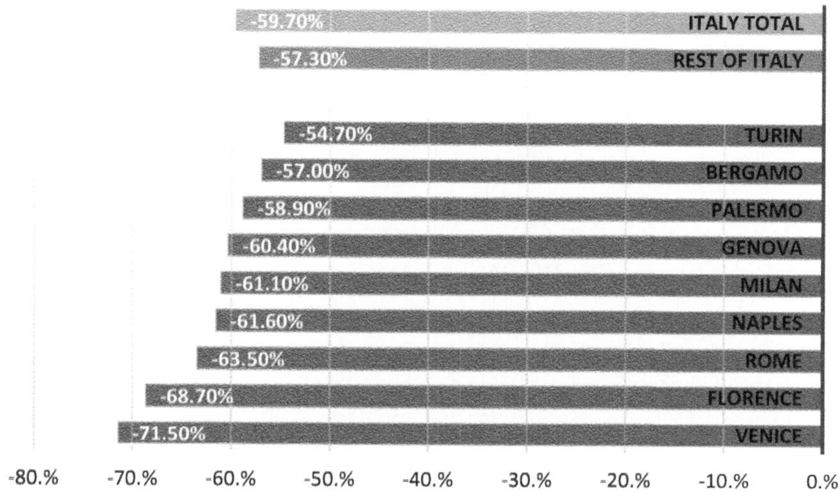

Figure 1. Travel to and Within Italian Cities in 2020.
Source: **Elaboration from National Tourism Agency (2021).**

make cities, and their tourism services, safer, greener, more innovative, intelligent, resilient, digital, and smart. As European Union tourism officials proposed the creation of a new Marshall Plan to finance a transition toward sustainability and adaptation to climate change, the last regime in order of presentation will take more space in the city and will need a strong cooperation with the actors of the other regimes to move in the European Green Deal context.

Chapter 4

SINO-RUSSIAN CROSS-BORDER TOURISM

Vladislav Chernov
Far Eastern State Transport University, Russia;
Pacific National University, Russia

Liubov Skavronskaya
Griffith College, Australia

Mariia Perelygina
Technological University Dublin, Ireland

Alexandra Bec
Griffith College, Australia; University of the Sunshine Coast, Australia

Elham Falatoonitoosi
University of the Sunshine Coast, Australia

Publisher's Note: Erratum
The publisher of 'Sino-Russian Cross-Border Tourism' in *Managing Destinations* has updated the author affiliation for Mariia Perelygina to read 'Technological University Dublin, Ireland'. This was due to a production error in the original incorrectly presenting the affiliation.

Managing Destinations: From Theories to Practices
Bridging Tourism Theory and Practice, Volume 14, 41–52

Published under exclusive licence by Emerald Publishing Limited
ISSN: 2042-1443/doi:10.1108/S2042-144320240000014004

Abstract: Geographic periphery regions are the focus of sustainable regional development research, with ongoing discussions on core–periphery relationships to address economic disadvantage and remoteness. However, periphery regions face challenges due to the high efficiency of core regions. Tourism provides an alternative economic landscape, revitalizing stagnant sectors and driving strategic exploration. The Sino-Russian border, spanning 4,209 km, represents a periphery where tourism plays a vital role. This chapter views Sino-Russian cross-border tourism through the lens of the core–periphery model, suggests two conceptual models of developing experiential cores in the cross-border areas of geographic peripheries, and argues for peripheries' potential to become sustainable experiential cores post COVID-19. **Keywords**: Sustainable tourism; Sino-Russian tourism; cross-border; core–periphery; COVID-19; tourism destination

INTRODUCTION

The potential for sustainable development and management of geographically remote tourism destinations is gaining attention in the recent tourism literature (Amoamo, 2018; Tambovceva, Atstaja, Tereshina, Uvarova, & Livina, 2020; Weaver, 2006, 2018). Although geographic borders are often viewed as obstacles for socioeconomic and sustainable regional development (Timothy & Tosun, 2003), a growing number of studies emphasize a positive impact resulting from cross-border tourism for the neighboring states. For example, regions may experience rich cultural exchanges, as well as significant economic benefits for local communities including increased social and infrastructure investment and an increased employment rate (Hampton, 2010; McKercher, 2001; Yuan, Fowler, Goh, & Lauderdale, 2013).

The Russian Federation is the largest country in the world, extending from Europe in the east to the furthest tip of Asia, sharing borders with several EUcountries (including Norway and Finland), China, Mongolia, Northern Korea, Japan, the United States, and former USSR countries, including Azerbaijan, the Ukraine, and Kazakhstan (Dimanche & Andrades, 2015). This empire developed through conquest of Siberia in the sixteenth and seventeenth century followed in the nineteenth century by an advance into the vast regions of the Amur and Ussuri Rivers, now known as the Far Eastern Federal District of Russia (Bassin, 1999). As a result, the Russian Federation today shares a 4,209-km border with the People's Republic of China (PRC) through quite remote territory. Most of this border is located between Heilongjiang Province in Northeast China and the Far East of Russia (Andrades & Dimanche, 2017). Although both countries meet at their peripheries, since the 1990s, trade initiatives and cultural exchange had thrived on the previously closed border and focused attention on the potential of cross-border tourism and sustainable regional development between China and Russia (Koziol, 2020; Telicheva & Chernov, 2020).

Large cross-border systems, such as the Sino-Russian border region, are characterized by mutual influence and integrity of two or more autonomous systems situated on opposite sides of the border, the intersection of the geopolitical interests of the neighboring states, and the asynchrony and asymmetry of changes on either side (Artemenko, 2010). The Sino-Russian relationship is of critical importance for political and economic stability not only in the Northeast Asian region but is also one of the most important bilateral relationships in the world (Hsu & Soong, 2014; Savic, 2016). In these circumstances, tourism plays a critical economic, social, cultural, and political role in the sustainable development of the region.

Despite its significance, studies on sustainable tourism management of cross-border destinations in the Russian Far East, especially published in English, are scarce. Specifically, the literature is void of studies considering cross-border Sino-Russian tourism using a core-periphery theoretical framework to investigate the region's potential for sustainable tourism development. In the recent literature, sustainable development of geographic periphery, cross-border regions are often contrasted with the development of spatial centers or "core" high-efficiency regions (Nuur & Laestadius, 2010; Robinson et al., 2014; Weaver, 2017). A periphery region is often a subject of negative perceptions for economic opportunity associated with social and economic disadvantages, remoteness, poor infrastructure, seasonality, and lack of resources, among a host of other challenges (Amoamo, 2018; Jewell et al., 2004).

In January 2020, Russia closed its Far Eastern border with China due to the spread of COVID-19 (TASS, 2020). On the April 24, 2020, the Russian government ceased issuing tourist visas to Chinese nationals, consequently stopping Sino-Russian tourism development (Ellyatt, 2020). This chapter is the first study that discusses cross-border sustainable tourism development opportunities in the Far East of Russia from core-periphery perspective, specifically exploring its potential to become an experiential core region. The following section focuses on historical and current aspects of tourism development in Russia and especially the Far East region.

Historical overview of tourism development in the Russian Far East

The first official evidence of tourism in Russia is from the end of the eighteenth century with far eastern travel beginning to appear in the Russian writing, literature, and culture in the middle of the nineteenth century (Chernov et al., 2019). Before the eighteenth century, there was no organized government regulatory system for managing tourism. Tourism development was characterized by individual initiatives, mainly the efforts of enthusiastic individuals pursuing religious purposes, trade, or political interests. The main motivators of Russian travelers in the nineteenth century were pilgrimage, medical tourism, education, and collection of geographic and historical information about various regions of Russia (Zorin, Kaverina, & Kvartalnov, 2005). Russia, in common with the rest of the western world, did not officially recognize tourism activities in the sense of a modern economic sector until the beginning of the twentieth century.

However, the location of the Far East of Russia between China, Korea, Japan, and the United States has meant that for centuries it has been a meeting ground for Eurasian and American economic and cultural exchange. Since the 16th-century reign of Ivan the Terrible, Russian leaders have sought to settle the Far East. However, the trade relationship between Russia and China developed greatly only after the *Treaty of Nerchinsk* (1689) was signed. Despite such progress, geographical remoteness was the main obstacle for regional development. For instance, in seventeenth century, a trade trip from Moscow to Beijing took approximately three years on average (Alexandrov, 1984).

Another obstacle for regional development of the Russian Far East is low population density. Since the Soviet Union's collapse in 1991, the region's population has fallen steadily. Russia's Far Eastern Federal District is two-thirds the size of China, however, the current population of just 6.4 million people and continuing to decline (Luxmoore, 2020). In contrast, the population of Chinese provinces across the Russian border, including Heilongjiang, Liaoning, Inner Mongolia and Jilin, etc is approximately 100 million people (Luxmoore, 2020).

The geographic position of Russia and the Far East inevitably led to the development of cross-border tourism: people living on the borders on both eastern and western civilizations expressed their interest in establishing trade connections and their curiosity to learn about neighboring cultures and new ways of life (Chernov et al., 2018; Dimanche & Andrades, 2015). Historically, this cross-border retail trade was concentrated in the large border cities of Blagoveshchensk, Khabarovsk, and Vladivostok. Mass cross-border tourism started after the building of the Chinese Eastern Railway (CER) (1897–1902) that connected the Russian Far East and Port Arthur, today part of Dalien in Liaoning Province, China. The railway was used by merchants, traders, and businesspeople from both sides. Several sanatoriums for workers and their families, as well as residents of Harbin and other cities were located along the CER. Harbin became a center of Russian culture and after the October Revolution in 1917, one of the centers of Russian emigration. There was a decline in relations between Soviet Russia and Chinese Manchuria during its occupation by Japan in the 1930s–1940s (Chernov et al., 2019).

After the Second World War, the Soviet Union provided economic assistance to China, and the first groups of Soviet tourists visited the country. Such group tours continued through 1956–1958 and were mainly taken by workers (coal miners, teachers, etc.) from the Russian Far East. Later, relations between the Union of Soviet Socialist Republics and the PRC deteriorated and tourist trips ceased. The resumption of cross-border tourism started only 30 years later when cross-border communication began by water transportation across the Amur River and regular flights between Khabarovsk and Harbin (Chernov, Kurbanova, & Zatesova, 2017).

A warming in relationships between Russia and China and significant development of cross-border tourism started in the early 1990s. The first governmental agreement on visa liberalization for tourist groups was signed in 1992. The development of trade relationships and the simplification of travel formalities for citizens of border regions prompted a sharp increase in shopping tourism (Perelygina, 2018). At the same time, territorial disputes between China and Russia significantly affected regional trade and cooperation for many years. A resolution of major parts of these territorial disputes was finalized in 2005, when Bolshoy Ussuriysky and Tarabarov Islands near Khabarovsk were ceded to China (Chow & Tsui, 2019). This territorial agreement prompted trade and tourism activities to develop, particularly along the cities near the Sino-Russian border. However, the COVID-19 pandemic since early 2020 has almost eliminated Sino-Russian cross-border tourism.

What Is Cross-border Tourism?

Cross-border tourism is a complex phenomenon with no agreed definition which is relatively unexplored in the tourism literature (Hampton, 2010). However, several studies propose that cross-border tourism predominantly involves day trip forays for leisure, business, visiting friends and relatives, or grocery shopping purposes, which involves crossing an international border (Chow & Tsui, 2019; Timothy & Butler, 1995). McKercher (2001) suggests that cross-border tourists exhibit significant differences in their travel patterns and prefer day trips with extended multi-night tours being rare. Compared with cross-border tourists, other tourists tend to stay longer and to include more stops on their travel itineraries. However, Hampton (2010) recognizes that the definition of cross-border tourism may include overnight stays, unlike conventionally defined international tourism.

A number of motivational factors for cross-border tourism are found in the recent tourism literature. Yuan et al. (2013) emphasize the role of functional and recreational shopping as one of the main motivations for cross-border tourism between Mexico and the United States, while Ryzhova and Ioffe (2009) suggest that Russian tourists from Blagoveshchensk are traveling to China's Heihe predominantly for recreational purposes, and some residents cross the Amur River every weekend. Nyíri and Breidenbach (2008) suggest that strong trade relationships between Russia and China, particularly for timber and minerals, have contributed to the development of the tourism sector. Thus, before the coronavirus global pandemic, trade and economic relations played a critical role in cross-border tourism development for the Russian Far East as both countries were receiving income from service provision, purchase of goods and souvenirs by business and leisure tourists. The economic importance of tourism for borderline regions of Russia and China is often a catalyst for other economic and infrastructural development (Hartman, 2016).

The coronavirus outbreak has revealed the vulnerability of cross-border tourism destinations and demonstrated that despite the decades of continual growth, the sustainable development of borderline regions is not guaranteed (Dedusenko & Wagenseil, 2020). The current destination management literature provides little guidance for a sustainable destination recovery post-pandemic (Galvani, Lew, & Perez, 2020; Moreno-Luna, Robina-Ramírez, Sánchez, & Castro-Serrano, 2021). Instead, academics and practitioners must search for "recovery road maps" (Dedusenko & Wagenseil, 2020). The following section introduces a strategy to develop an experiential core on a geographical periphery.

An Experiential Core at the Geographical Periphery

The concept of a core-periphery is used in many contexts but is commonly applied in geographic models to describe areas of differing economic production and political power (Hall, Harrison, Weaver, & Wall, 2013). Simply, it's a concept used to understand disequilibrium among regional economies and providing a hierarchical structure across economic systems (Amoamo, 2018). In a geographical context, a core-periphery model describes dependency relationships between a dense core region and an unconnected periphery region (Robinson et al., 2014). In an economic context, the model explains unequal structural relations, such as size or composition, between the advanced or metropolitan "center" and a less developed "periphery." From a broader philosophical perspective, core-periphery represents an analytical framework to understand economic dualities, such as strong-weak, change-stability, etc. (Hall et al., 2013; Mahutga, Ma, Smith, & Timberlake, 2010). In a tourism context, a core-periphery framework has been applied to understand the dynamics of global, regional (Weaver, 2017), and rural-urban destination development (Amoamo, 2018), as well as the potential of geographic peripheral destinations to provide peak experiences (Pegas, Weaver, & Castley, 2015; Weaver, 2006).

As mentioned previously, geographical peripherality is often associated with negative social and economic growth perceptions. Jewell et al. (2004) categorized some of these negative perceptions as low levels of economic vitality, dependency and declining traditional industries, geographical remoteness, lack of technology and innovation, poor infrastructure, difficult visitor access, seasonality, lack of education, resources, and experiences. The idea of an experiential core in a geographic periphery explains how a remote economically marginal destination can be modified to meet the needs of tourists and become a center for growth in provision of tourism experiences (Turner & Ash, 1975; Weaver, 2017). Hall (2006) highlights two key elements characterizing an experiential core on the geographical periphery; (1) a pleasant landscape and (2) the temporary flow of people between core regions and the tourism periphery. Thus, providing opportunities for developing sustainable economic systems in periphery regions.

The current tourism literature stresses the paradoxical nature of the core-periphery phenomenon in the context of geographically remote destinations, where destinations have opportunities to become an experiential core (Chaperon & Bramwell, 2013; Robinson et al., 2014; Weaver, 2006, 2017). Figure 1 demonstrates traditional core-periphery model, as well as a

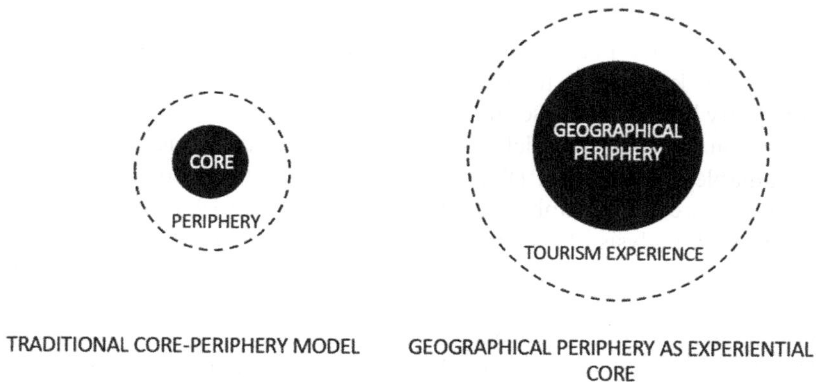

Figure 1. Geographical Periphery as an Experiential Core (Proposed by the Authors).

paradoxical core–periphery relationship when a geographically remote tourism destination becomes an experiential core.

The tourism literature acknowledges the positive influence of an experiential core on sustainable regional development at a peripheral tourism destination (Lee, Weaver, & Prebensen, 2017; Weaver, 2006, 2017). Thus, tourism has a capacity to facilitate sustainable regional development through transforming a geographical periphery into experiential core which then subsequently influences growth of other economic sectors. Additionally, the paradoxical nature of core-periphery relationships in a tourism context aligns with current arguments for adopting a paradox lens for tourism research (Sigala, 2020). In addition, current COVID-consumer tourism studies suggest that the focus of destination management initiatives must be promotion to local tourism markets, as well as on strengthening collaboration (Zhang & Blasco, 2021). Therefore, cross-border areas not only have the potential to become sustainable experiential cores in the post COVID-19 era but also represent the areas where a post pandemic tourism recovery will start from, with the Far East of Russia representing one of these strategic cross-border areas.

Sustainable management destination management in cross-border areas

Sustainable destination development

Sustainable tourism is a complex phenomenon that includes strategic planning, tourism management and marketing initiatives, and measuring, monitoring and evaluating tourism activities at different levels. Weaver (2017) suggests that there is a minimalist and a comprehensive composite model of sustainable tourism development, and argues that whichever model is adopted, effective planning and management decisions required relevant and adequate knowledge to address the complexity of the situation. The sustainability literature stresses the need for any tourism organization to have a clear vision and goals that define that vision, as well as a practical definition of sustainable development (Weaver, 2006). Thus, to build a sustainable experiential core at geographic peripheries post COVID-19, a comprehensive inventory and analysis of the strengths and weaknesses of the destination is required, following the development of a clear vision and sustainable development goals. This forms the starting position for new development and improvements (Dedusenko & Wagenseil, 2020).

The 2030 Agenda for Sustainable Development (United Nations, n.d.) emphasizes the importance of partnerships and strategic alliances and calls for action by all countries to work together to end poverty, promote prosperity (social and economic), and protect the environment (Falatoonitoosi, Schaffer, & Kerr, 2021). Sustainable development requires understanding of the interactions between politics, environment, economy, and culture, and how they influence environmental sustainability, social inclusion, and overall prosperity (Falatoonitoosi et al., 2021).

Development of sustainable tourism at geographic peripheries in the post COVID

Peace, safety, security, and political stability are fundamental requirements for tourism development success at the destination level (Dwyer, Edwards, Mistilis, Roman, & Scott, 2009). To avoid popular overcrowded destinations, unconventional and emerging destinations will be in high demand and their development will make tourism more sustainable (WTTC, 2020). Even before COVID-19, there was a growing tourist demand toward more meaningful, memorable, and emotional tourism experiences at idyllic and isolated destinations (Moyle, Moyle, Bec, & Scott, 2019; Robinson et al., 2014; Skavronskaya, Moyle, Scott, & Kralj, 2020; Weaver, 2017). Due to the COVID-19 pandemic, the tourism community must adapt to a new reality of social distancing and other associated interventions intended to prevent the spread of the coronavirus disease. To restore cross-border Sino-Russian tourism destinations post pandemic, a holistic sustainable tourism development strategy specifically tailored to the needs of the region is required to provide an opportunity for destination management organizations in both countries to build experiential cores for cross-border geographical peripheries on the Sino-Russian border.

Maintaining and managing a COVID-19 safe tourism destination is easier on the geographical periphery, compared to core tourism destinations. On peripheral destinations, such as, for instance, the Far East of Russia, it is easier to trace contacts, maintain social distancing, respond to potential outbreaks, and maintain hygiene practices. However, maintaining COVID-19 safety in a tourism destination is not enough to restore and reset the tourism industry in cross-border areas. A holistic sustainable tourism development strategy is also required.

Experiential cores at Sino-Russian geographical peripheries: A core-periphery approach

An experiential core on a geographic periphery is the place where tourist flows from the core engage with attractions (Lacey, Weiler, & Peel, 2014). The existing literature suggests that tourists experience a destination through hedonic consumption – the act of pleasure (Weaver, 2018). Therefore, it is important to understand tourists' motivation, goals, decision-making process, and challenges in the post-COVID-19 era to be able to design the experiences with an opportunity to transform them into experiential cores. Based on the traditional core-periphery model (Figure 1), this chapter suggests alternative variations for developing experiential cores in the cross-border areas of geographic peripheries (Figure 2).

Since the beginning of Sino-Russian tourism development, both countries have made multiple attempts to establish tourism models for destination development on cross-border areas. The most successful example in the Russian Far East is the "Twin Cities" of Blagoveshchensk (Russia) and Heihe (China). The "Twin Cities" model is demonstrated in Figure 2a. The successful tourism development between Blagoveshchensk and Heihe is determined by geographical proximity, visa-free initiatives, and transport accessibility (Mikhailova, 2015).

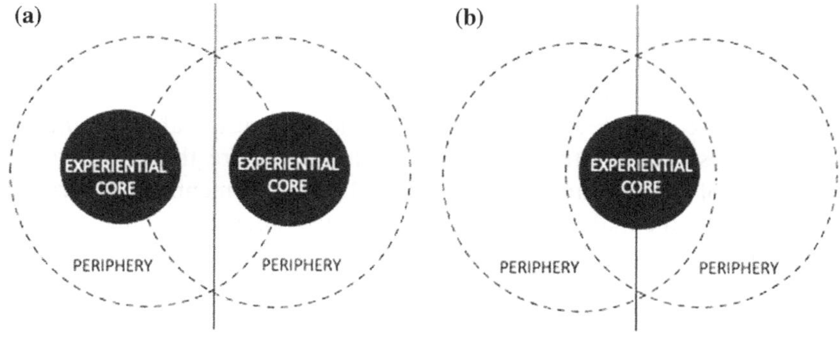

Figure 2. Conceptualizing Models of Experiential Core Development in a Cross-border Geographic Periphery (Proposed by Authors).

There is potential for other peripheral cross-border regions to become experiential cores. For instance, Bolshoy Ussuriysky Island (Chinese: Heixiazi) is an island shared between Russia and China and therefore provides potential to become a unique experiential core (Figure 2b). Potentially, the shared island can become a destination for short tours and getaways for residents from both countries. Especially, the introduction of new relaxed visa policies or even a visa-free regime on the island for citizens of both countries would facilitate the development of this tourism cluster and increase the numbers of visitors for both sides. The development of touristic and transportation infrastructure also requires collaborative efforts. At the moment, China has developed transportation communication on its side (tourist boats and road and rail service from the town of Fuyuan) and a nature reserve with basic touristic infrastructure including a wetland park with a wooden walkway, a botanical garden, and a nine-story pagoda pointing China's most eastern territory (Vasyukova, 2018). While currently the Russian side remains relatively unused and underdeveloped (a part is used for agriculture/farming), projects of the development of the Russian side of Bolshoy Ussuriysky Island suggest creating a join resort area with hotels, business centers, shopping malls, luxury apartments with potential facilities of an aquapark, a golf course or an equestrian club. Another opportunity is turning this space into a gambling zone. One more project suggests the addition of cruise tours along the Amur River to the local tourism cluster. Given the unique environmental assets of this area and the preservation of the untouched nature on both sides of island, there are opportunities for dark sky tourism, off-road tourism (ATV & Buggy tours, snowmobiles, dog sledding), wildlife tourism including birdwatching, fishing, and hunting. The initial developments of the nature reserve on the Chinese side may also be extended to the Russian side.

The rare case of sharing the island creates wide opportunities for cooperative development and building a single touristic complex with complementary tourism offerings from both countries. To achieve this aim, both countries need to combine their effort to build a strategic partnership to restore Sino-Russian tourism while also considering sustainability principles. This includes collaborative investment in infrastructure that promotes the uniqueness of the geographic regions, as well as complements the synergies in existing and potential tourism behavior. Such alignment can foster the development of a tourism hub for specific types of tourism activity, promoting strategic future tourism investment and sustainable long-term planning.

Overall, we argue that applying a core–periphery lens to Sino-Russian cross-border tourism is an innovative way to reconceptualize and restore the destination image, trade, cultural relationship, and tourism between two countries in the future post-pandemic world.

CONCLUSION

The tourism literature provides several approaches for managing tourism destinations in cross-border areas with studies suggesting focusing on assessing political and economic factors, sociocultural interactions, and government legislation (Kozak & Buhalis, 2019). For instance, Baggio (2008) argued that a cross-border system can be examined through its socioeconomic potential, geopolitical significance, or individual features of the border. Many studies have focused on economic, political, and cultural comparative analysis of the border areas (e.g., Hampton, 2010; Mikhailova, Wu, & Chubarov, 2018). Recent studies suggest that the border regions, such as the Far East of Russia, have significant potential for sustainable regional development through tourism (Dunets, Ivanova, & Poltarykhin, 2019).

The Sino-Russian cross-border system is a borderline system where both countries adapt to each other through tourism, economic, and cultural exchange. This study recommends that Russian Far East regions develop a new vision, goals, and objectives for their tourism policies and planning that align with sustainable development. In this chapter we explored how a core-periphery theoretical model can be applied to the cross-border tourism context to further investigate the potential for sustainable development. The core-periphery model originates from a general systems theory and is not new for tourism research. In a geographical context, a core-periphery is an analytical framework used to understand domination – dependency relationships between a dense core and unconnected periphery (Robinson et al., 2014). A growing number of studies emphasizes paradoxical nature of the core-periphery relationships tourism context. The paradox of a core-periphery analysis is that peripherality can equally be considered as a core from the perspective of a tourism experience. Thus, geographically remote destinations have an opportunity to become an experiential core and to assist in overcoming inequitable core-periphery relationships, therefore providing an innovative approach to sustainable regional development through tourism (Lee et al., 2017; Pegas et al., 2015; Weaver, 2006, 2017, 2018). In this chapter, we applied the model to Sino-Russian cross-border tourism in the Far East of Russia.

Sustainable regional development in cross-border areas, such as the Far East of Russia, remains challenging (Collins & Cunningham, 2017). Geographical remoteness, lack of funding, limited economic resources, poor tourism infrastructure, and the global COVID-19 pandemic are only few factors significantly affecting the cross-border tourism between Russia and China. The impact of the COVID-19 on the travel industry is yet to fully unfold (Nicola et al., 2020) and the recovery of the global tourism industry from it depends on the typology, scale, and scope of the crisis (Matiza, 2020; Ritchie & Jiang, 2019). Although the severity of COVID-19 has limited international and, in some cases, domestic tourism, its impact can be mitigated by avoiding overcrowding in tourist accretions. Uncrowded tourist destinations also align with the major premises of sustainable tourism development and can help in the sustenance of the tourism industry post-COVID-19.

Considering this argument, as well as the existing regulations on lockdown rules and social distancing, it is suggested that sustainable tourism practices can become a potential solution to stimulate tourist movements and help the revival of the tourism industry in cross-border areas, such as the Far East of Russia. This makes more sense than ever before, as it has been also argued that sustainable tourism development can become part of a solution to the crisis caused by COVID-19, with the core–periphery model providing a powerful framework with which to address sustainability, as well as safety issues. Under a core–periphery model, tourists from the core engage with peripheral experiential cores in geographically remote destinations, where COVID-19 safety regulations are easy to maintain and control. Therefore, we recommend considering sustainability practices in tourism planning post pandemic not only because of greater environmental awareness but also because of the obligations that will be imposed on travel and choosing a destination, including peripheral tourist destinations.

Existent research showed that tourists have high intentions to select sustainable touristic destinations. Therefore, we recommend that enhancing sustainability practices in remote destinations such as Sino-Russian cross-border areas can reposition sustainable tourism development while focusing on its pre-existing competitive advantages post-pandemic.

Sustainable tourism development has been also considered as an alternative to more traditional means of poverty alleviation (e.g., Cobbinah et al., 2015). According to 17 Sustainable Development Goals, all countries need to work together to end poverty, promote prosperity, and protect the planet, which requires an understanding of the interactions between politics, environment, economy, and culture, and how they influence environmental sustainability and social inclusion post-COVID-19 (United Nations, n.d.)

Overall, this chapter provides a first study exploring Sino-Russian cross-border tourism through the lens of a core–periphery framework highlighting opportunities for sustainable regional development, post-COVID-19 era. Future studies can explore the opportunities to transform other geographically remote areas to become experiential cores and their potential to elicit peak experience, such as, for instance, Russian Arctic Tourism, which was gaining popularity among Chinese tourists before the pandemic.

Chapter 5

CULTURE-LED REGENERATION Mestre, Italy

Olga Tzatzadaki
University IUAV of Venice, Italy

Abstract: This chapter discusses how an urban regeneration process, with culture as an economic asset, can contribute to creating a city brand and identity for the city of Mestre, a new habitat, and economic opportunities for the local community, as well as helping neighboring Venice tackle overtourism. This chapter highlights that places suffering from undertourism often are located closely to those suffering from overtourism. Urban regeneration policies driven by culture can be a key solution for both places, by creating a new future, identity, and economic opportunities for one community and for the other, in helping to regulate their touristic flows. **Keywords**: Culture-led; urban regeneration; city brand; undertourism; city marketing; Venice

INTRODUCTION

This chapter proposes a culture-led regeneration plan for Mestre, Italy. Mestre is characterized by undertourism, and the policy recommendations of this plan will benefit it by creating its own urban identity. It will also work in favor of Venice, by helping the latter in regulating its tourism. First, this chapter provides a theoretical discussion of culture-led regeneration processes, their birth and causes in the deindustrialized Western world, of the post-Fordist era. Next, the birth and evolution of overtourism and undertourism and their impact on local communities are discussed. The second part provides a case study of Mestre and its relation to Venice, and research conducted in the Piave district. Data were collected through site inspections with further graphical information system mapping of the economic activities, accommodation, and real estate statistics from the Agenzia delle

Managing Destinations: From Theories to Practices
Bridging Tourism Theory and Practice, Volume 14, 53–66

Published under exclusive licence by Emerald Publishing Limited
ISSN: 2042-1443/doi:10.1108/S2042-144320240000014005

Entrate, Real Estate Observatory 2010–2017. Nonstructured interviews were undertaken with a random sample of residents and local shop owners, and a content analysis was conducted of the last 5 years of media reports. The results are presented in a Strengths, Weaknesses, Opportunities, and Threats (SWOT) analysis, which summarizes the local economical and socio-spatial profile as well as its main characteristics (Table 1).

Given the strong relationship of the local community with culture and art, a culture-led regeneration process is proposed as a solution to the problems identified. Through this process, Mestre will benefit from a new urban identity differentiated from Venice. By becoming the contemporary alter ego of Venice (all year round but also during the *Venice Biennale* following the example of Zona Tortona in Milan), Mestre will offer residents and tourists all that Venice cannot. This is a youthful, contemporary urban core, in which creativity, art, fashion, and culture can blossom, converting the city into a must-see and must-be-seen-in destination.

Deindustrialization and Culture-led Regeneration

From the beginning of the 1970s, cities in the Western world experienced de-industrialization and the relocation of their industries to countries with lower labor costs such as China and East Europe, centralization of control functions, and the evolution of technology used in the production processes. As a result, cities have had to reinvent themselves by elaborating their self-image to continue to attract investments and people. In this process, the role of Italian cities was strengthened by administrative decentralization, which allowed them to become financially autonomous.

Over time, cities gained a strategic role in the global arena, bringing an increase in international trade and foreign direct investment (Moulaert, Rodriguez, & Swyngedouw, 2005; Sassen, 1991). Cities and ex-industrial cities of the Western world tried to create a new identify for themselves. Cities promoted new images, concentrated in specialized knowledge, dissemination and interpretation of information, know-how, innovation and their competitive advantages, facilities and, in general, in converting themselves into service-oriented economies. Urban planning played a key role in deciding why, where, and how these changes would take place, and in the process included private actors in the planning. Instead of regulating urban growth, planning had to encourage it through regeneration processes and changing the city's image, its brand.

This transformation of industrial cities into service-oriented economies was accompanied by a growing interest in using culture as a tool for urban regeneration (Garcia, 2004). Here urban regeneration is defined as a form of comprehensive management intended to continuously improve economically, socially, and physically declined areas via partnerships between the public and civil sectors on a strategic plan for the progress of the whole city

(Pinfield, 1997). Evans and Shaw (2004) suggest that regeneration is the transformation of a place that shows symptoms of environmental (physical), social, and economic decline or, better, as the infusion of new vitality to communities, industries, and places in decline, bringing sustainable and long-term improvements in economic, social, and environmental terms.

It was American cities that made the first experiments in the 1970s, while it was after the 1980s that Europe also began to consider culture as a tool for urban regeneration. The works of Scott (2000), Evans (2001), Byrne (2002), and Florida (2002) on culture-led development explain in detail this approach. Although cities have always had cultural functions, it was the first time that culture was considered as an economic asset, a commodity with market value and, as such, a valuable producer of marketable city spaces. Culture was used to help reinvent cities as centers of excellence for business and tourism consumption. Richard Florida (2002) in his famous work, *The Rise of the Creative Class*, proposed that the presence of a creative class, a class of entrepreneurs, cultural operators, administrators, etc., was indispensable to ensure the rebirth of cities in decline due to causes related to deindustrialization.

Today, culture-led regeneration has become a key aspiration for most cities and regions, particularly those having suffered the effects of post-industrial economic decline (Garcia, 2007). In this process of urban regeneration, culture is a driver, a catalyst, or at the very least, a key player. The main goal is the consideration of culture not as a temporary tool toward external ends, but as an end in itself, which will develop its full potential in the long term. A regeneration process driven by cultural activity emphasizes identity issues, as well as creative developments, often linked to artistic creation. The benefits commonly associated with culture-led regeneration initiatives are the strengthening of local identity and self-confidence, and the ability to attract and retain creative talent (Garcia, 2007).

Kyu Hong Hwang discusses the connection of culture and urban identity "By transforming a low-value and deteriorated area into a highly valued district, the local community can simultaneously restore its social pride, revive the local economy, and realize an urban identity" (2014, p. 67). Landry (2018) considers heritage and creativity are great partners, but a rich cultural tradition makes someone less innovative because of the tendency to maintain what is achieved so far. Landry considers the key is to shift ambition and understand that you must be creative today to survive in the global stage of economy. Cultural activities, both traditional and new, create meaning and thus are concerned with and embody the identity and values of the city, both in terms of what it was and is becoming.

A brilliant example of a culture-led regeneration process can be found in the Docklands of East London, a former industrial area, abandoned and in transition (Landry, 2018). Dockland's low land value attracted young artists in a moment when, because of the lack of interest in investing in the area, it was affordable for them and had lots of immigrant communities. With time, Docklands started to grow and to develop different projects. Individual artists and small micro-businesses began to aggregate. Today, some of the most important companies in the world are in this zone. The power of creative businesses transformed the area. There are open markets, but also consumption-oriented facilities, many different organizations, and companies, which all create an interesting economy and place-to-be. As a result, the area became a touristic attraction, transforming it into not only a great place to live but also a great place to visit. A collaboration between people and businesses of different fields generated innovation, a key ingredient for all cities which managed to survive the deindustrialization process. Other examples of culture-led regeneration include Bilbao, Barcelona, Glasgow, Liverpool, Bristol, Newcastle-Gateshead, Marseilles, and Leipzig, and Ruhr.

In Italy, zona Tortona of Milan provides an instructive example. This was an industrial district in the south of Milan, created at the end of the 19th century. After the abandonment of the factories and between the 1980s and 2000s, the area experienced a transition from a production site for heavy industry to a creative hub for fashion, art, cinema, photography, and design. This ex-industrial district is now the center of production, no longer material, but immaterial, and a place of innovation. In the area, a new residentiality is present, which penetrates the urban landscape of the original and surviving local fabric. The *Fuorisalone* Fair and Fashion Weeks are hosted in zona Tortona each year, filling its streets with thousands of tourists. They are attracted because of the events, and for the creative and joyful atmosphere that the area offers, transforming it into a must-see for tourists and for the residents of Milan.

Nevertheless, all best practice urban regeneration processes require collaboration between the public and private sector, in other words, between all actors and stakeholders interested in the project. To achieve the best possible results, public administrators, private investors, residents, and urban regeneration actors should collaborate. Top-down policies are necessary for the project to succeed and to be coordinated and at the same time, bottom-up action and private investment are indispensable. There is no ideal project design for all contexts (one-size-fits-all approach). Instead, all ideas, proposals, and policies should be calibrated to local needs and that is the reason why a local socio-spatial profile is critical for a project to succeed. Studying local socio-spatial features is necessary to create tailor-made urban policies, which will suit the local context and community and make a project work. Without collaboration with the local community, the project is likely to fail. Understanding local communities' needs is crucial for the success of urban policies.

Overtourim and Undertourism

Koens, Postma, and Papp (2018) provide an interesting overview of the overtourism literature. Until recently, tourism was seen as one of the most sustainable economic growth strategies for cities. Particularly, in the aftermath of the economic crisis of 2008, it was viewed as an important driver for economic recovery or growth and was given plenty of scope to develop, thus reinforcing its relative importance in city destinations (Russo & Scarnato, 2018). However, the perception of city tourism has changed dramatically. Public transportation, infrastructure, roads, museums, attractions, and other services that were primarily created for local use are suffering under increasing tourist numbers. The growing popularity of online accommodation services (AirBnB, HomeAway, Uber) and a desire to see real and authentic everyday city life has meant that tourism activities become intertwined with local life, even outside of the main tourism areas in cities (Pappalepore, Maitland & Smith, 2014). Such developments have led to an increasingly vocal call from residents and local stakeholders to deal with tourism growth, and protests have been observed in multiple high-profile destinations such as Barcelona and Venice. Although the issue is most prominent in European cities, similar sentiments have been reported in other destinations too, for example, tropical islands, backpacker ghettos, or slums (Koens & Thomas, 2016). The term overtourism is used to describe these tourism disturbances.

The term overtourism arose from media discourse. The issue it describes, an excessive negative impact of tourism on the host communities and/or natural environment, has been a concern within academia for many years. Forster (1964) and Wagar (1964) mentioned its negative impacts on destinations and later Doxey (1975), Pizam (1978), and Butler (1980) described the social cost for local communities. In the 1980s and 1990s, scholars discussed the carrying capacity of destinations (McCool & Lime, 2001; Van der Borg, Costa, & Gotti, 1996). Later, tourist behavior, timing, concentration, and location, as well as tourist numbers were discussed (Lindberg, McCool, & Stankey, 1997; Postma, 2013). Most recently, academics have concentrated on how to manage the negative impact of tourism through government policy.

Overtourism began as a term referring to the overuse of natural resources (Nelson, 2002; Nguyen & Shadrin, 2008). Gerosa Bellows (2012) made the term popular, using it in the *National Geographic Travel* journal. The United Nations World Tourism Organization defines overtourism as "the impact of tourism on a destination, or parts thereof, that excessively influences perceived quality of life of citizens and/or quality of tourists' experiences in a negative way" (2018, p. 4). Koens, Postma, and Papp (2018) indicate that current academic work is focused on the relation between tourism, its wider city context, and the political aspects of excessive tourism growth, mentioning the works of Milano (2018), Freytag and Bauder (2018), Smith et al. (2018), and Shoval (2018). Overtourism involves the interplay of tourism and urban change and the need for regulation and government leadership (García-Hernández, De la Calle-Vaquero, Yubero, & García-Hernández, 2017; Nofre, Giordano, Eldridge, Martins, & Sequera, 2018; Novy, 2018; Postma, Buda, & Gugerell, 2017).

The term undertourism is more recent and was listed as a travel megatrend for 2019 (Peltier, 2019). Edgell Sr (2020) suggests undertourism concerns emerging or less visited destinations that can be an alternative to overcrowded places. These may be underrepresented destinations or with few tourists during the off season and in need of tourism spending. As Kennel notes (2018), most destinations are struggling with undertourism. In developing countries, attracting tourists is a means to economic and social development. Kennel (2018) suggests that the main causes of undertourism are a lack of government attention and funding, poor destination management, and increased competition. As a result, undertourism often is associated with unemployment, abandonment, social and economic problems, in other words in a general deterioration of the socio-spatial context. In parallel, Peltier notes that "Amsterdam, Barcelona, Venice, and Kyoto still contain unknown treasures and unbeaten paths if you know where to look and when to go. The answer for these and other destinations is not to undersell because of overtourism but to adopt smart strategies for marketing the alternatives that tourists did not know they loved" (2019, n.p.). These are destinations in need of a strategic marketing plan to become an appealing touristic destination and, most importantly, to survive economically. Marketing policies for destinations affected by undertourism requires creating a destination image, or else a well-defined self-brand, to differentiate themselves from other touristic offers. Urban planning and policies can contribute to reaching this goal.

Destinations affected by overtourism often tend to overshadow their neighboring destinations which, in many cases, could be the key for managing the massive touristic flows of popular places. In this case, shifting attention toward these nearby destinations by creating the right policies can reduce overcrowding. Tourism can offer a new vision and opportunities for these overshadowed destinations. In the long term, the result is the creation of a great place to live and a great place to visit.

CULTURE-LED DEVELOPMENT FOR MESTRE

To understand the urban context of this case study, a historical overview is necessary. According to the Venice Municipality Archives, in 1926, the Municipality of Mestre was merged with Venice, within the framework of a national administrative rationalization. This entailed the elimination of several municipalities throughout Italy. On this occasion, all the smaller neighboring municipalities (Burano, Chirignago, Favaro Veneto, Mestre, Pellestrina, Zelarino) were merged into Venice. This also happened in all the Italian provincial capitals. At the time, Venice had a population of just over 200,000 inhabitants, compared to 20,000 in Mestre. After the Second World War, with the arrival of the "Italian economic miracle," all urban centers, especially in the North, underwent urban and economic development, which led to the formation of suburbs around the historical town centers.

The urbanization process was caused by the transition (also found in the Veneto region) from an agricultural to an industrial society, which entailed the need to build new residential neighborhoods near industrial areas. An increasing number of Italians aspired to live in innovative, reinforced concrete houses, which at the time were considered modern and healthy (with heating, running water, electricity, and sanitation). In comparison, traditional rural houses were dilapidated and lacking in services. Unfortunately, this urbanization process was poorly planned, with few rules to regulate and orientate the process. Land speculation was common and there was little respect for the environment or historical monuments. Starting from the 1970s, the deteriorating historical centers of medium and large cities and municipalities in Northern Italy became depopulated.

In Venice, urban development mainly took place on its mainland, as the building space in the lagoon was limited, especially in the historic center. Mestre, therefore, welcomed Venetian urban development and from a small peripheral center of 20,000 inhabitants, it became a suburb of Venice with a population of 200,000 people. However, unlike a normal suburb, Mestre was not incorporated into the development of the center and remained distinct from it. Mestre was more accessible than the Islands of Venice. Consequently, today's Mestre is a peripheral urban center, but also a more modern part of Venice.

In comparison, at the beginning of the 20th century, Venice was unable to become an industrial and port city capable of competing with the other centers of the Mediterranean due to the lack of a suitable land. At the same time, a demographic problem arose, as the historic center of Venice did not surplus building space and most of the inhabitants were crammed into overcrowded houses with poor hygiene (Tagliabue, 2012). Coen Cagli addressed these problems in three ways: the construction of a commercial port (350 ha, to which was added a small 30 ha oil port), an industrial area (700 ha), and a residential district (225 ha) in the Marghera district, in the southern part of Mestre. The planning for the Marghera industrial area began in 1919, and the construction of the residential district began on May 10, 1921.

In the 1990s, Venice's administration wanted to develop tourism, and urban policies were proposed to achieve this. At the same time, Mestre, which served as the residential area of Marghera and Venice, was left without a clear vision of its own and without clear functions. Today, Mestre is dependent on Venice's city brand and touristic profile and is seen as a cheaper accommodation alternative for tourists. Venice receives more than half of all tourists to the region with 9.5 million arrivals in 2017 (out of a total of 19 million in the province) and 37 million visits (out of 69 overall). Of these 37 million, more than 11 million were to Venice city with 7.8 million to the historic center, 3.2 million to Mestre, and 0.5 million to the Lido (Italian Ministry of Economy and Finance, 2017).

Table 1. SWOT Analysis of Piave District.

Strengths	Weaknesses	Opportunities	Threats
1. Transportation Accessibility through train, tram, and bus Proximity to Venice and Mestre city center	*1. Economy* Lack of facilities for tourists High homogeneity of economic activities	*1. Directly related* Recent touristic investments Opportunities for co-working	*1. Directly related* Increasing usage of airport Administration is oriented toward Venice's benefits rather than Mestre's
2. Space Available spaces for investments Low land value	*2. People* Lack of connections between locals and immigrants No places for young people to interact	*2. Larger economic environment* Increasing GDP percentage of service in Veneto region Increasing tourism in Venice	*2. Larger economic environment* Low-added value of service Similar economic activities with Venice
3. People Large and diverse population Many students and young people in the city	*3. Government* Insecurity and distrust to the government and local administration		

The Piave District

Mestre's urban functions and services are connected to Venice and to the industrial district of Marghera, and it lacks a brand of its own. Mestre constitutes an ideal candidate for a culture-led regeneration process. The goal of this research was to select an area within Mestre suitable for an urban regeneration process. The Piave district was chosen as it has several useful characteristics. The Via Piave is a one-mile-long road which connects Mestre's railway station and the city's center, Piazza Ferretto. According to locals, in the 1980s and 1990s, the Piave district was one of the most expensive streets in terms of land value, with prestigious retail shops and apartments. In the last 15 years, due to socioeconomic changes, it is now considered as one of the most degraded areas of Mestre, with low land values. Residents avoid the area. Piave Street is a mixed-use area, with commercial activities, residential apartments, offices, and warehouses.

Specific research questions were formulated for study of this area. What is the area's commercial profile? What is the connection between tourism and the economic activities of the area? What services are provided to residents or to tourists and how are they distributed? How does the security problem affect economic activities? Is there space for new investments and what are the land values of it?

The area under study is close to the city center, and Mestre Railway Station, one of the biggest and most important railway stations of Italy, with thousands of commuters each day. Venice is 10 minutes by train and 15 minutes by bus and Marco Polo Airport, one of the most important airports of the country is nearby. The area contains Piraghetto Park, one of the city's biggest and most visited park by the residents, and the VEGA research park and start-up incubator, which hosts hundreds of workers and students every day. There are abandoned warehouses near the railway station. These spatial characteristics constitute a series of significant socio-spatial conditions that will support redevelopment.

Several different approaches and methodologies were used to collect data: graphical information system elaboration and mapping of economic activities and accommodation in the selected area and their characteristics. Interviews were conducted with residents and local shop owners, and real estate statistics and media reports were reviewed. Site inspections identified 445 commercial and 117 accommodation businesses including hotels, and bed and breakfast, and Airbnb facilities.

Maps were created showing the types of economic activities (markets, entertainment, restoration, services), customers (residents, tourists, mixed use), business closing times (before or after 7:00 p.m.), open and closed business, and the distribution of accommodation facilities.

Only six out of a total of 445 businesses identified were for entertainment (1%). Commercial activities were orientated toward local customers and only 10% (45 units) were for tourists and closed later than 7.00 p.m. Nearly a quarter of the commercial spaces had closed (22% or 129 units). In the accommodation sector, 84% of the businesses (98) are Airbnb apartments, while only 16% (19 units) are formal hotels.

A content analysis of media articles of the last 5 years indicated security problems and reports of drug distribution, while most of the recent stories (during the last 2 years) discussed issues related to new accommodation buildings in the railway area. These were low-cost accommodation and hostels capable of hosting about 2000 people per day when finished in 2020. Average land values had fallen by around one third between 2010 and 2017, and many buildings are for sale (*Agenzia delle Entrate*, Real Estate Observatory 2010–2017).

Residents, shop owners, students, and commuters were interviewed to collect data on the daily life, needs, fears, and desires of the local community. Shop owners wanted more touristic flows and customers, while for residents, the local market profile and products do not match their needs. In addition, locals complained about the lack of entertainment in the evening. Foreigner-owned businesses were more adaptive and optimistic than Italians, while Italian owners seemed nostalgic of the past and believed that the extra-EU community is the main reason of the degradation of the area.

Data collected indicates showed that the area has low land values and is suitable for investments, such as that already taking place in the touristic sector. In the area, nearly one third of people are foreigners, in other words, it is a multicultural local community. In Mestre, there is a large student community because of the lower rents and the universities located in Venice and Mestre. This means that there is the need of more activities for younger people and related to culture and art (Table 1).

Mestre's Vision

Currently, traditional Venetian souvenirs, like masks and *Murano* glass, are produced in Eastern countries at low cost and imported to Venice. These are sold to low-or-no-budget tourists, who constitute close to 70% of Venice's arrivals. Tourists can find hundreds of shops in Venice that sell the same products at a fifth of the cost of authentic ones. This, along with the enormous high rents for commercial activities in Venice, are the main reasons why artisans are forced to close. Gianni De Checchi, the secretary of *Confartigianato*, an organization that assists small businesses and artisanal activities, notes that in 1966 there were 2,800 artisanal companies in Venice, while today they are 1,800. The loss of a strong urban economy is a challenge for Venice. A strong urban economy is constructed by diversity in opportunities in the professional field for its inhabitants and a network of activities and people. Because of the economic crisis, more and more young people have an interest in traditional artisanal activities. This is an attractive solution to unemployment compared to other saturated professions.

Moreover, interest in craftsmanship and artistic activities is written in the locals' DNA. Local urban policies and territorial marketing should be based on these specific socio-spatial characteristics. This will lead to promotion that is embraced by locals and which has a positive impact on the community. The Venetian region has been for centuries famous for its artisanal products and Venetian craftsmanship and art is a part of its identity. The relationship between Venetian craftsmanship and art was, according to Giuseppe Mazzariol (Zanzotto, 1992), one of the greatest creative and economic drivers of the Venetian territory. Today, many artisans cannot afford the high rents for their studios in Venice. They are few and dispersed, so the chances to create partnerships and networks are low.

The proposals for Mestre and the Piave district are based on a culture-led regeneration plan, with art as a trigger for development. The intention is not to copy Venice's artistic and cultural profile, but to build its contemporary alter ego in Mestre. What Venice is missing is a contemporary urban core suitable for the younger generations to interact and to live in. There are many Venetian universities in the city with thousands of students. Moreover, Venice's tourists need a lively place for amusement and entertainment to visit in the evening. Venice does not offer such a place.

These considerations are the basis for the proposed policies. The Piave district could be the ideal place for young artists' studios and Venetian artisanal laboratories. The empty and rent affordable spaces of the Piave district are suitable for coworking spaces and everyday interaction and artistic inspiration. Moreover, the Piave district could host a more experimental, alternative artistic event during the yearly Venice Biennale exhibition of Art and Architecture with a younger and more experimental profile. One example of this proposal is traceable in Milano's Zona Tortona, the *Fuorisalone* event, which takes place every year during the international Furniture and Design Fair, the *Salone del Mobile*. Zona Tortona has experienced a culture-led regeneration and during the *Salone del Mobile* fair each year, an alternative artistic exhibition (called *Fuorisalone*) takes place (differentiated in content from the *Salone del Mobile*). During this exhibition, zona Tortona is visited by thousands of people and during the rest of the year, it remains one of the most interesting parts of the city for young people to interact, to work, and to live in. Alternative theater venues play to small audiences and open concerts, festivals of ethnic gastronomy, and open markets help to activate the area. Venice needs a similar place where contemporaneity can be welcomed, and this could be Mestre's Piave district.

Mestre needs to provide more than inexpensive accommodation in comparison to Venice. It should offer different activities to Venice's tourists. Mestre could target its own types and develop its own identity. With time, Mestre will then attract more tourists as well as new inhabitants, will create small businesses, and a place where young people can interact, have fun, and create networks and partnerships.

These policy recommendations can help to regenerate the area, offering young artists and artisans a place to relocate with low costs, create activities for tourists, and an attractive place to live. This will eventually allow Mestre to compete with Venice in a different way, offering its residents and tourists all that Venice is lacking. Shifting a part of Venice's touristic flows to Mestre will also help Venice in tackling the overtourism issue. With time, a culture-led regeneration development will create an interesting habitat; a nice place to visit and a nice place to live in.

As noted above, culture-led development can succeed, no matter what the scale of the city. As Charles Landry (2018) highlights, it is the era of the "second cities" because sometimes, big creative cities become unaffordable. In this project, the collaboration between the local administration, the local community, and private businesses is crucial to create synergistic networks. The administration should help in orienting the development and facilitate bureaucratic procedures, while artists should work together, reinforcing each other and creating networks. The aggregation of *creativity* in a geographical area can generate innovation and help Mestre gain a clear identity and an urban economy of its own. This is important for Venice, and for the whole region.

Nevertheless, culture-led development in Mestre is not only about triggering the economy indirectly and giving a brand to the city. There is also a sociological factor based on relatively new research that shows that the participation in, and consumption of, cultural activities can contribute to (2019, n.p.) well-being. Research by Grossi, Sacco, Tavano Blessi, and Cerutti (2011) studied Italian residents and showed that culture has a relevant role as a determinant of individual psychological well-being. These researchers found that a selected subset of cultural variables are the best predictors of individual psychological well-being. Individual well-being is substantially affected by specific forms of cultural access, and therefore, policies aimed at fostering cultural access can be regarded (and suitably reframed and redesigned) as health policies. According to this investigation, city planning based on culture is of sociological importance, as it is related to public health, well-being, and happiness. Venice is the city center of the Veneto Region with thousands of commuters traveling to it every day for work and study. Mestre could be that part of the city center that Venice is not. It could be a more informal, active area, in which contemporaneity as well as diversity can be welcome. A culture-led plan for the city could contribute to changing locals' lives, economically and socially, offering them a better quality of life and a city image, with benefits for themselves and for Venice as well.

CONCLUSION

Each socio-spatial context needs urban policies calibrated to its profile, strengths, and weaknesses. According to this tailor-made approach, policies which are related to overtourism or undertourism issues cannot be applied in the same way in all contexts suffering from their effects. The Venice case deals with an entire city, not with parts of the city only, as in other contexts. This means that policies which propose gates or entrances to the city will eventually transform Venice into a theme park. On the contrary, Venice needs policies, which tackle the problem indirectly through transversal policies, which offer multiple benefits and resolve it from the inside. Mestre's role, in this direction, could be crucial for Venice. Enhancing Mestre's identity and differentiating it from Venice could be lifesaving not only for Mestre itself but also for Venice.

Destinations suffering an undertourism effect need a marketing plan and the creation of a destination image. It is recommended that urban policy proposals for Mestre should start with a culture-led regeneration process. Using culture and art could work as a trigger for the economy, offering Mestre an identity and differentiated touristic appeal from that of Venice, helping the latter by shifting a great part of its flows. The creation of a contemporary artistic core in the Piave district through artisanal laboratories, open markets, and multiple cultural events throughout the year as well as hosting the alternative, more experimental face of the *Biennale Exhibition* of Venice could eventually help Mestre build a strong city image. These policies, calibrated to local society's needs, are an answer to both Venice's overtourism and Mestre's undertourism. The strategic position of Piave district, easily reachable from Venice and Mestre's railway station and from the city center, its low land values, and multicultural community are the ingredients for the success of such policies. With time, more and more services will aggregate together, attracting more tourists, who will then attract more businesses, giving birth to a series of benefits for the local community. Furthermore, it will offer activities also for Venice's residents and tourists during the day and in the evening for the younger generations to interact.

Moreover, the Veneto region needs a coherent plan for dealing with *Venice's overtourism* effect and for improving the general touristic offer. Within the same region, overtourism coexists with a strong undertourism effect, which are both risky for its communities. Until now, solutions for those two contexts, Venice and Mestre, seem to be not in a dialogue with each other and in communication with a greater vision for the future and the development of the Region. Collaboration between the multiple destinations of the Veneto region like Venice, the UNESCO'S Prosecco Vineyards, the Palladian Villas, the historic city center, and the Dolomiti UNESCO site located in the Belluno Province is necessary. Overcrowded sites need their neighbor areas as allies and partners.

As already noted, perhaps undertourism, which certainly concerns more destinations than those concerned by overtourism, is in fact more dangerous than overtourism. Unemployment, abandonment, and social, cultural, and economic difficulties – all effects of undertourism – concern numerous communities and these locations are often situated close to the ones suffering from overtourism. A strategic marketing plan concerning both locations is indispensable, as both destinations need the other.

Collaboration between all actors concerned in these territories is necessary. Decisions about such transformations should include all voices present in the territory to guarantee a high quality of life and economic opportunities for the residents of the territory. The success of top-down policies is impossible without the collaboration of all actors present in the territory and, at the same time, bottom-up operations cannot succeed without the contribution of the administration. An equal distribution of the multiple benefits of tourism, among neighbor territories, is necessary to tackle overtourism and undertourism affected territories.

Acknowledgments—The research group involved in this project included Israel Goldshied, Kristen Liang, Olga Tzatzadaki, Michiel Van Baelen, and Jessica Wang, University of Venice. While the study was conducted by the group, the culture-led policies elaborated in this chapter belong to the author.

NEW MARKETS

Chapter 6

MADEIRA TOURISTS PROFILE
A Positive Psychology Approach

Soraia Garcês
University of Madeira, Portugal

Margarida Pocinho
University of Madeira, Portugal

Saúl Neves de Jesus
University of Algarve, Portugal

Abstract: Today's tourists seek authenticity and high-quality experiences. This chapter presents exploratory results from a tourist well-being project conducted in Madeira Island, Portugal. Using a tourism well-being scale, onsite and online data were collected in 2019. The study suggests that tourists with higher well-being are younger, have lower qualifications, are retired and/or students, first timers, and mainly Portuguese (mainland). Participants identify gastronomy, walk in nature, and landscape as the three most enjoyed activities/experiences. Results showed that tourists in Madeira want to have fun, experience something unique, and develop positive relationships. This study opens doors to offer customized experiences considering tourists psychological profile and their well-being. **Keywords**: Psychological tourist profile; well-being; activities; experiences; Madeira Island; positive psychology

INTRODUCTION

Today, maybe more than ever before, tourism is changing. This change is leading people to look and to choose the place they want to visit and spend

Managing Destinations: From Theories to Practices
Bridging Tourism Theory and Practice, Volume 14, 69–83

ISSN: 2042-1443/doi:10.1108/S2042-144320240000014006

their money carefully and differently. Europe is the lead destination in the world having reported in 2019 an increase of about 4% in international arrivals when compared with the same period the year before (European Travel Commission [ETC], 2019a). Thus, tourism is a fundamental part of Europe' economy, society, and culture (ETC, 2018).

Tourism is a part of the experience economy (Kirillova, Lehto, & Cai, 2016; Pine & Gilmore, 1998, 2019) and means a shift in European strategic marketing (ETC, 2019b). Tourists want to be intrinsically connected with destination experiences (Busser & Shulga, 2018). This connection can be developed through cocreation, leading to memorable and indeed transformative experiences for the tourist (Campos, Mendes, Oom do Valle, & Scott, 2016; Kirillova, Lehto, & Cai, 2017). Memorable experiences have in satisfaction with the destination a mediator variable regarding revisitation intention and "positive word-of-mouth" (Gohary, Pourazizi, Madani, & Chan, 2018). Thus, destination policies are important supporters since hospitality, staff attitude, hotel rooms, location, and food can contribute to these memorable experiences (Sthapit, 2018).

People are increasingly valuing the idea of quality of life, meaningful and fulfilled experiences. Thus, the links between tourism, well-being and life quality are growing (Uysal, Sirgy, Woo, & Kim, 2015). The number of tourists looking for experiences that result in well-being is rising. Thus, well-being can be a predictor of place attachment (Vada, Prentice, & Hsiao, 2019). New well-being products are increasingly popular, especially those offering eudemonic experiences (Smith & Diekmann, 2017). All tourists are not alike, and "their intentions and desires, needs and wants are changing" (Garcês, Pocinho, & Jesus, 2019, p. 101). Today's tourists are more spontaneous, increasing search for authenticity and quality of experiences, have a new care for the environment and sustainability, and are more aware of local cultures (Paulisic, Tankovic, & Hrvatin, 2016). Experience quality can act as a mediator of authenticity and satisfaction when tourists visit cultural heritage destinations (Domínguez-Quintero, González-Rodríguez, & Paddison, 2018). This suggests that an authentic and high-quality experience increases tourists' satisfaction with cultural tourism activities. Hence, a focus on destination activities that lead people to feel good is a potential growth spot for tourism (Sangpikul, 2010).

A strategic transformation in the new European program, Horizon 2022, highlights experiences and human behavior as central parts for tourism development in the coming years (ETC, 2019b). The Portuguese 2027 tourism strategy recognizes the central position of people in tourism endeavors. Individuals are "transversal assets" in all activities (Turismo de Portugal, 2017). The country has also highlighted well-being as part of its 2027 strategy considering it an "emergent asset" (Turismo de Portugal, 2017).

Positive Psychology is a field that explores how people and communities thrive (Seligman, 2013). Well-being (in its many forms) is a central concept of the field. Its models have emerged with the positive emotions, engagement, relationships, meaning, and accomplishment model (PERMA) (Seligman, 2011). The character strengths and virtues model (Peterson & Seligman, 2004) is based on six virtues (each with strengths) found in almost all cultures: wisdom and knowledge (creativity, curiosity, open-mindedness, love of learning, perspective), courage (bravery, persistence, integrity, vitality), humanity (love, kindness, social intelligence), justice (citizenship, fairness, leadership), temperance (forgiveness, modesty, prudence, self-regulation), and transcendence (appreciation of beauty and excellence, gratitude, hope humor, spirituality) (Peterson & Seligman, 2004). Applying Positive Psychology in tourism can help the industry to understand tourism experiences (Dan, 2015). PERMA can be aligned with the tourist experience in a destination. For example,

> When daydreaming about the upcoming vacation our positive emotions might increase, …when mindfully proceeding some of the environmental or architectural details of sites we visit we feel engaged, …when laughing and joking with travel companions or host communities' bond and build relationships, … achievement as another outcome of importance might be a priority after bungee jumping or white-water rafting, …when we witness extreme poverty on our trips, we may question our meaning or sense of purpose in life. (Filep & Pearce, 2013, p. 227)

Madeira Tourism

The tourist well-being project applies positive psychology to the study of well-being in tourism, implemented in Madeira Island, Portugal. The project "aims to bring new and much-needed data about tourists' wellbeing, to increase knowledge about the tourism phenomenon, characterize the tourists' profile, and ultimately, lead to tourism products and services requalification" (Garcês et al., 2019, p. 111). The project connects the PERMA model with three other variables: creativity, optimism, and spirituality. There are few studies about optimism in tourism. Yet, this variable is related to well-being and satisfaction important to tourism stakeholders (Garcês, Pocinho, & Jesus, 2018a). Creativity is usually studied from the perspective of a place and its culture and not from tourists' perspectives. Tourism can be considered a spiritual activity connected to personal fulfillment and so with the idea of "experience economy." Creativity, optimism, and spirituality are important positive psychology outcomes.

Tourism experiences that allow a tourist to see situations (even bad ones) positively, that are unique and original, and that encourage a search for meaning provide new ideas and possibilities that can contribute to the competitiveness of the industry and the study of well-being (Garcês et al., 2018a). Thus, the operationalization of the tourist well-being project began by linking PERMA with creativity, optimism, and spirituality. This led to a theoretical model: human optimal psychological experience model. This is an interactive and dynamic model for well-being and the core of the tourist well-being project (Garcês, Pocinho, & Jesus, 2017). It is defined as

> ...a personal state of fulfilment that occurs during an "experience" in which the person's wellbeing suffers influences from his/her creativity, optimism, and spirituality. As in a loop situation, individuals' wellbeing influences their "experiences" in a given moment and is influenced by them. Moreover, their creativity, optimism, and spirituality can affect the activities an individual enrolls during an "experience" and therefore it will affect their wellbeing. (Garcês et al., 2017, p. 340)

The second stage of the project started with data collection using the tourism well-being scale (Garcês, Pocinho, & Jesus, 2018b, 2019, 2020a). The project is currently in the stage of data analysis. Next will come the development of an app to allow customization of experiences.

Madeira Island is one of the oldest destinations in the world (Almeida, 2016). It has climate as a major attractive offer (Marujo, 2013). This destination has won the Europe's Leading Island Destination title six times and the World's Leading Island Destination title five times. Tourism is responsible for about 15% of the region employment, and it is its main economic activity (Secretaria Regional de Turismo, 2016). The Madeira tourist profile provides the economic and demographic characterization of tourists. Tourists in Madeira mostly arrive through Madeira Airport, led by Portuguese, German, English, and French tourists. They are likely married, with a high school or a university degree and an annual income above US$14,646 (Secretaria Regional de Turismo, 2016). Madeira's tourism strategy is positioned on a cluster of characteristics that make it a unique place to enjoy nature, year-round mild climate, high-quality service, tradition/hospitality, history and culture, diverse sea and mountain activities, markets proximity accessibility, and safety and security. Its products are divided into three main categories: strategic, central to the tourists' experience (Nature, Sports Nature and Adventure, Cultural and Landscape Touring, Nautical and Sports, Sun and Sea); complementary, not central but important to the overall experience (health and well-being, gastronomy, and Madeira wine); and secondary, existing on a small scale (meetings, golf, resort, and residential) (Secretaria Regional de Turismo, 2016).

An earlier study found Madeira's meaningful experiences have three main focus: nature, culture/heritage, and economy/marketing. A lack of emphasis regarding meaningful experiences is clear (Garcês, Pocinho, & Jesus, 2020b). Therefore, the tourist well-being project is a first attempt to study well-being and meaningful experiences in Madeira Island, to look for its potential as an innovative and creative product for this destination. This chapter offers the first analysis of the empirical results of the project. It explores the psychological profile of Madeira tourists and the activities/experiences that bring them well-being in respect to their individuals' profile.

MADEIRA TOURISTS' PSYCHOLOGICAL PROFILE

The sample consists of 471 tourists, who were randomly selected through social media and in the city streets. Ages ranged from 18 to 84 ($M = 41.7$, SD $= 17.6$). Female made for 59.8% of the sample ($n = 282$) and male for 40.2% ($n = 189$). The sample is composed of 65.6% active workers ($n = 309$) and 26.1% of nonactive workers (students and retired people) ($n = 123$). Almost half of the participants have high school completed or less ($n = 226$; 48.0%), and the remaining have a bachelor's or a higher degree ($n = 218$; 46.3%). About 41.4% ($n = 195$) of the tourists had already visited Madeira at least one time (repeaters). The participants visited Madeira between February and December of 2019 from 37 countries. The majority came from the mainland Portugal ($n = 102$; 21.7%), Germany ($n = 84$; 17.8%), the United Kingdom ($n = 77$; 16.3%), Czech Republic ($n = 61$; 13%), and rest ($n = 147$; 31.2%) from other countries.

The instrument used was the tourism well-being scale (TWS) (Garcês et al., 2020a) and sociodemographic questions. Assessing tourists' well-being, the instrument is comprised of eight items that measure well-being (positive emotions, engagement, relationships, meaning, and accomplishment), creativity, optimism, and spirituality. Although it is a unidimensional measure, it can examine the items individually. The validation study applied exploratory and confirmatory factor analysis and displayed acceptable psychometric results. Responses are given using a Likert scale, varying from totally disagree (1) to totally agree (7). Reliability was 0.87, and it showed convergent validity. In the current study, overall reliability is 0.83.

Authorization to collect data in the streets was obtained through the City Hall of Funchal, and then a team of volunteers started data collection. Data were collected in two ways: in the streets, through random and direct contact in the main city of Madeira, Funchal; and through social media with the dissemination of a Google Form. Ethical standards about confidentiality and anonymity were clarified to participants. Participants were adults ($> = 18$ years old) and had at least one tourism experience in Madeira Island. Data collection took about a year. Statistical software Statistical Package for the Social Sciences (SPSS, version 26.0) was used for descriptive, correlational, and inferential statistical analysis, including the analysis of equality of variances through Levene's test. Content analysis was performed for the open question regarding Madeira Island's most enjoyed activities/experiences and a coding frame created. As recommended, the triangulation of multiple sources of evidence (quantitative and qualitative data) was performed to obtain a psychological profile of the Madeira tourist (Yin, 2003).

Study Results

Measures of central tendency were calculated to summarize the data for the study variables. Measures of dispersion were computed to understand the variability of scores for these variables. The global score of TWS was a mean of 43.7 (SD = 7.29), with a minimum of 12 and a maximum of 56. Most tourists' responses were at the Likert scale level four "Agree in Part" considering the seven Likert levels. The ranking of the eight items of the TWS is presented in Table 1. The first chosen item was positive emotions, the second creativity, and the third relationships. The item with less punctuation was spirituality. The Pearson *r* correlation coefficient was computed to assess the relationship between each of the eight items. There was a strong positive correlation between all items of TWS [$0.194 < r < 0.548$, $p < 0.001$].

Independent-samples *t*-test was performed to compare well-being TWS items for males and females. There are no statistically significant differences regarding gender for all the analyzed items ($p > 0.05$). Independent-samples *t*-test was used also to analyze well-being according to tourists age groups: a younger group of participants with less than 40 years old and an older one with subjects 40 years old or more. Considering age and the total score of the

Table 1. TWS Items Ranking ($N = 471$).

Rank	Description	Variables	*M*	SD
1	I had lots of fun.	Positive emotions	6.09	1.07
2	I faced this experience as a unique/ original opportunity.	Creativity	5.77	1.27
3	I felt good in the relationship I developed with new people.	Relationships	5.68	1.26
4	I engaged in the community activities (ex. cultural, events, etc.).	Engagement	5.59	1.38
5	I was able to see the positive side of the less agreeable situations that occurred.	Optimism	5.30	1.26
6	This experience was a dream come true.	Accomplishment	5.23	1.50
7	I found out new ways of being that gave meaning to aspects of my life.	Meaning	5.18	1.35
8	I experienced a connection/relationship with something higher than myself.	Spirituality	4.85	1.73

TWS, there was a significant difference between younger participants (M = 44.69, SD = 6.49) and older (M = 42.58, SD = 7.96) participants (t (469) = 3.17, $p < 0.001$). These results suggest that younger tourists showed higher well-being than the older ones. Analyzing the items of TWS one by one, results showed the existence of significant age differences only in four variables: positive emotions, creativity, achievement, and relationships, with younger tourists performing better than older tourists. The effect sizes, considering means, standard deviation, and sample size, Cohen d and Hedges' g are small-moderate for the variables' relationships and achievement and for the TWS global score. For creativity and positive emotions, the effect size is moderate (Table 2).

Independent-samples t-test was performed to compare well-being according to tourists' qualifications groups: a group of participants with

Table 2. TWS Age Differences.

TWS Items	Variables	Age Group (Years Old)	n	M (SD)	p	Cohen d/ Hedges' g
I had lots of fun.	Positive emotions	<40	309	6.33 (0.87)	0.000	0.45
		> = 40	123	5.83 (1.20)		
I faced this experience as a unique/original opportunity.	Creativity	<40	309	6.02 (1.02)	0.000	0.42/ 0.43
		> = 40	123	5.49 (1.45)		
This experience was a dream come true.	Achievement	<40	309	5.40 (1.45)	0.010	0.24
		> = 40	123	5.04 (1.53)		
I felt good in the relationship I developed with new people.	Relationships	<40	309	5.82 (1.13)	0.015	0.23
		> = 40	123	5.53 (1.37)		
TWS global score	Tourist well-being	<40	309	44.69 (6.49)	0.002	0.29
		> = 40	123	42.58 (7.96)		

12th grade or less and another one with subjects with higher education (bachelor's or higher). Considering education level and the total score of the TWS, there was a significant difference between participants with a high school degree or less (M = 44.54, SD = 6.55) and tourists with higher education (M = 43.15, SD = 7.54; t (442) = 2.079, $p < 0.039$). Analyzing the items of TWS individually, results showed the existence of significant differences regarding qualifications, only in the achievement and in the spirituality variables. Less qualified tourists performed better in both variables when compared with who had higher qualifications. The effect sizes, considering means, standard deviation, and sample size, Cohen *d* and Hedges' *g* are small (Table 3).

Regarding the job situation (active and nonactive worker), significant differences were found. In the global score of TWS, there was a significant difference between tourist active workers (M = 43.43, SD = 7.29) and tourists with a nonactive job situation (M = 45.34, SD = 5.83; t (471) = −2.864, p = 0.004).

The effect size, considering means, standard deviation, and sample size, Cohen *d* and Hedges' *g* are small-moderate. These results suggest that having a job (active worker) or not (nonactive worker – retired or student) influences tourists' well-being. Results suggest that nonactive workers' well-being is

Table 3. TWS Qualifications Differences.

TWS Items	Variables	Qualifications Level	*n*	*M* (SD)	*p*	Cohen *d*/ Hedges' *g*
This experience was a dream come true.	Achievement	< = Secondary	226	5.39 (1.43)	0.026	0.21
		> = Bachelor's	218	5.08 (1.51)		
I experienced a connection/ relationship with something higher than myself.	Spirituality	< = Secondary	226	5.03 (1.70)	0.049	0.19
		> = Bachelor's	218	4.70 (1.76)		
TWS global score	Tourist well-being	< = Secondary	226	44.54 (6.55)	0.039	0.19
		> = Bachelor's	218	43.15 (7.54)		

higher than the active workers' tourists. Analyzing the items of the TWS individually, the study results showed the existence of significant job differences only in four variables: creativity, engagement, achievement, and spirituality. Nonactive workers performed better than active worker's tourists (Table 4).

From the total of the sample (N = 471), 41.4% (n = 195) were repeat tourists. This study focused on the last visit to Madeira or the present visit, in the case of the face-to-face data collection surveys. Analyzing the items of TWS individually, significant differences emerged between first-timer tourists and repeat tourists. These differences were found in three variables: meaning, creativity, and achievement. First-timer tourists performed higher than those who already had visited Madeira Island (Table 5).

Table 4. TWS Job Situation Differences.

TWS Items	Variables	Job Situation	n	M (SD)	p	Cohen d/ Hedges' g
I faced this experience as a unique/original opportunity.	Creativity	Active	309	5.73 (1.24)	0.002	0.29/ 0.31
		Non-active	123	6.09 (0.98)		
I engaged in the community activities (ex. cultural, events, etc.).	Engagement	Active	309	5.52 (1.34)	0.013	0.27
		Non-active	123	5.88 (1.35)		
This experience was a dream come true.	Achievement	Active	309	5.15 (1.49)	0.012	0.28/ 0.27
		Non-active	123	5.54 (1.30)		
I experienced a connection/ relationship with something higher than myself.	Spirituality	Active	309	4.76 (1.76)	0.011	0.28/ 0.27
		Non-active	123	5.23 (1.60)		
TWS global score	Tourist well-being	Active	309	43.43 (7.29)	0.004	0.29/ 0.28
		Non-active	123	45.34 (5.83)		

Table 5. First-Timers and Repeat Tourists Differences.

TWS Items	Variables	Tourists	*n*	*M* (SD)	*p*	Cohen *d*/ Hedges' *g*
I found out new ways of being that gave meaning to aspects of my life.	Meaning	1st timer	272	5.29 (1.32)	0.045	0.19
		Repeat	195	5.04 (1.36)		
I faced this experience as a unique/original opportunity.	Creativity	1st timer	272	5.95 (1.32)	0.001	0.29
		Repeat	195	5.55 (1.43)		
This experience was a dream come true.	Achievement	1st timer	272	5.37 (1.38)	0.026	0.21
		Repeat	195	5.06 (1.61)		
TWS global score	Tourist well-being	1st timer	272	44.52 (6.74)	0.007	0.25
		Repeat	195	42.69 (7.83)		

In the global score of TWS, there was also significant differences regarding first-time tourists (M = 44.52, SD = 6.74) and repeaters (M = 42.69, SD = 7.83; t (465) = 2.009, p = 0.007). These results suggest that the first visit to Madeira had a higher effect on tourists' well-being than the subsequent visits. The effect size of first-timer and repeater differences is small, based on means, standard deviation, and sample size of the variables meaning and achievement, Cohen *d* and Hedges' *g*. For creativity and TWS total, Cohen *d* and Hedges' *g* are small-moderate.

One-way ANOVA was computed to analyze psychological profiles differences by countries. The results displayed significant differences for three variables: meaning [F(3) = 2.749; p = 0.043], creativity [F(3) = 4.815; p = 0.003], and relationships [F(3) = 4.131; p = 0.007]. A *post hoc* Bonferroni test was computed to assess multiple comparisons. Portuguese tourists performed better than Germans in meaning (p = 0.027), creativity (p = 0.031), and relationships (p = 0.019). Czech Republic tourist's results showed higher means differences than Germans' in creativity (p = 0.012). The United Kingdom tourists performed better than German's in relationships (p = 0.017). The ANOVA effect size for the variable meaning is small

(η_2 = 0.02). But it is small-moderate for creativity (η_2 = 0.044) and relationships (η_2 = 0.037).

A qualitative content analysis method was used to analyze the qualitative data collected through the open question: "What activities/experiences did you most enjoy?". A coding frame was created, followed by a segmentation of the material. Then, a trial of the coding frame occurred, followed by its evaluation and finally by carrying out the main coding. The following 13 categories were achieved: Levada's walks, landscape, nature, walking, flowers and gardens, hiking; mountains, gastronomy (food, drink, and restaurants), sea; people and hospitality, cultural events (festivities, Madeira's lifestyle), cities and villages, cable car, the Monte sledges, and island trips. The coding frame criterion was the first activity/experience participants answered to the above open question. It is possible to see (Table 6) that the activity/experience mostly enjoyed was gastronomy, followed by Levada's walks and landscape.

Table 6. TWS Ranking Items by Each Activity/Experience.

Activities/Experiences		TWS Items Ranking								
	n	1	2	3	4	5	6	7	8	TWS Global *M*
Levada's walk	**50**	PE	C	R	E	M	O	A	S	41.42
Landscape	**43**	PE	R	C	E	A	O	M	S	41.58
Nature	39	PE	C	R	E	O	S	A	M	43.89
Walking	26	R	PE	E	C	M	A	O	S	43.58
Flowers and Gardens	17	E	PE	R	C	O	S	A	M	42.50
Hiking	17	PE	E	C	A	R	O	M	S	44.47
Mountains	18	PE	C	R	E	O	A	M	S	44.22
Gastronomy (food, drink, and restaurants)	**52**	PE	C	E	R	M	A	O	S	45.38
Sea	34	PE	C	R	E	O	A	M	S	43.74
People and hospitality	23	R	PE	C	O	E	M	A	S	47.04
Cultural events (festivities, nigh life)	40	PE	E	C	R	O	A	S	M	44.9
Cities and villages	30	PE	C	E	O	R	A	M	S	42.8
Cable car, Monte' sledges, Island trips	22	PE	C	R	A	M	O	E	S	44.8
Total*	436									

Notes: M = Meaning, PE = Positive Emotions, O = optimism, C=Creativity, E = Engagement, A = Accomplishment, S=Spirituality, R = Relationship, * "Other activities" were not count on this table because the *n* was less than 5 cases.

The bold highlights the three activities/experiences with higher frequencies.

Table 6 provides a first psychological profile of the Madeira tourist. This tourist is someone who seeks/has positive emotions, unique and original experiences/opportunities, positive relationships, engagement in local activities, optimist, achievement, meaning in their lives, and spiritual.

Psychological Profile of Tourists

This chapter analyzes the empirical results of the tourist well-being project developed for Madeira Island, Portugal. It aims to explore the psychological profile of the Madeira tourist and the activities/experiences that enhance well-being regarding their individual' profile.

Overall, results showed that there are no gender differences. But age had an effect on well-being experiences. Younger tourists showed higher well-being experiences than older tourists, positive emotions, creativity, achievement, and relationships. Tourists with fewer qualifications performed better in the achievement, spirituality, and overall well-being. Nonactive workers performed better than active ones in global tourist well-being (TWS total) and, specifically, in creativity, engagement, achievement, and spirituality variables. Furthermore, significant differences emerged between tourists who visited Madeira Island for the first time (first-timers) and repeaters. These differences were found in the TWS total score, meaning, creativity, and achievement variables, with the first-timers performing higher than those who already had visited Madeira Island. Analyzing psychological profile differences by the four most frequent countries, Portuguese tourists performed better than Germans in meaning, creativity, and relationships. Czech Republic results were higher than the Germans in creativity. The United Kingdom performed better than Germans in the relationship variable.

Regarding tourists' psychological profiles based on their well-being experiences, data showed that positive emotions was the most frequent variable being in the first place for 10 in the 13 activities/experiences mentioned by tourists. The exceptions were only for activities/experiences such as walking and people and hospitality, with the variable relationship being in the first place. This suggests that tourists felt good in the relationship they developed with new people during the walks and in the interaction with local people. In the same way, another exception is the flower and garden activity with the engagement variable being in the first place. This possibly means that tourists were more engaged with the Madeiran community traditional events, like flower festival and gardens visits (botanic gardens, for example). Creativity was the psychological variable in second place present in 7 of the 13 activities most enjoyed by tourists. Creativity is related to Levada's walk, nature, mountains, gastronomy (food, drink, and restaurants), sea, cities, and villages; Cable, the Monte sledges, and island trips. These results suggest that participants found these experiences as unique/original opportunities. In third place comes relationships, in six of the 13 activities. This is the most frequent item in the third place, which suggests tourists felt good in the relationship they developed with new people during the activities/experiences: Levada' walks, nature, flowers and gardens, mountains, gastronomy (food, drink, and restaurants), sea, cable car, Monte sledges, and island trips.

Reflecting on the study findings and current literature, differences between older and younger tourists in travel lifestyles, motivation to travel, and activities engaged while on vacation were also found in other study (Acevedo & Nohara, 2004). Others showed that senior tourists have different needs (Vigolo, Simeonu, Cassia, & Ugolini, 2018) than younger ones.

Concerning the relevance of relationships, in another study regarding the motivation for tourists to take part in volunteer or humanitarian activities, four main factors were found: immersing in the locals' culture, "to give back," building friendships, and family (Brown, 2008). Other studies found three major subjects regarding destination image: natural resources, built resources, and people (McCreary, Seekamp, Davenport, & Smith, 2019). Others found out that the main motivators for creative activities in the context of creative tourism involved "sightseeing, social contact, self-improvement and escape" (Zhang & Xie, 2018). These results reflect the importance of the relationships developed with locals and is in accordance with the current study, leading to believe that positive relationships are relevant for tourists' well-being.

While in the current study, positive emotions were the most frequent variable, others have found that negative emotions can contribute to eudaemonic experiences as well (Nawijn & Biran, 2018). The study also found that the experiences engaged by the tourists are diverse, but almost all highlight Madeira's unique culture and landscape. Other researchers have found that it is possible for tourists to experience a destination through food, culture, and traditions (Boonpienpon, Maneenetr, Siriwong, & Kovathanaul, 2015).

A particular relevant finding of the current research is the fact that spirituality was almost always in the last place. However, other studies acknowledge that different kinds of spiritual tourism activities are increasing (Cheer, Belhassen, & Kujaw, 2017). Through tourism experiences, it is possible to comprehend meaning, transcendence, and relationships with others (Willson, McIntosh, & Zahra, 2013). Studies have also found that traveling can lead to seniors' spirituality growth (Moal-Ulvoas, 2017). Thus, tourism can be regarded as a spiritual endeavor connected to values and a search for personal and social development (Singh, 2009). However, this means that in this study, spirituality is not yet seen as a driver for Madeira tourism.

Overall, tourists have an array of motivations to travel. But they are looking for authenticity, for personal development, and enjoyment. Madeira tourists want to have fun, to have unique experiences, to interact, and develop relationships with new people and to engage in local activities. They are also able to see bad situations as temporary and consider the experience a dream come true. Finally, the experience enriched their lives, and it allowed to connect with something higher than themselves, even in a small scale.

CONCLUSION

On Madeira, tourists with higher well-being can be characterized as being 40 years old or less, with fewer qualifications, retired and/or students, who have visited Madeira Island for the first time, and come mainly from Portugal mainland, followed by Germany and England. Results suggest that an optimal experience in Madeira includes positive emotions, creativity, and relationships. The experiences that promoted more well-being were the ones where tourists connected with people (hospitality). Participants identify gastronomy, Levada's walks, and landscape as the three most enjoyable activities.

Overall, the activities/experiences on the island contribute to tourists' well-being. The tourist profiles can be used to offer customized experiences based on tourists' psychological characteristics and adjust current product offers and develop new marketing strategies. This study describes tourists, not in an economic view, which is the usual one, but from a social and a psychological perspective. It is a very challenging endeavor and ambitious, much needed outlook which can help tourism thrive and survive in this destination.

Nevertheless, every research has its limitations. In this study, the number of participants (N = 471) is not representative of the Madeiran tourist population *per* year (1.4 million) (Secretaria Regional da Economia, Turismo e Cultura, 2016). The sample should be expanded in future research. Another possible improvement is to conduct interviews with tourists to get deeper and richer psychological data. The study provides valuable insights into what is a significantly under-researched topic about the interconnection between psychological well-being profiles and tourist activities/experiences. The psychological characteristics of the main profiles of Madeiran tourists support future sustainable growth on the island.

Policymakers should consider these findings in their decision-making and tourism activities planning and improve current policies. Future research should examine less enjoyable activities on the island (crowds of tourists, dangerous roads and transportations, expensive flights) and their implications for government policy. It is crucial to debate the possibility of developing more sustainable management strategies that minimize the negative impact of tourism and maximize the tourist and residents' well-being in destinations.

Individuals choose a destination for many different reasons. "Some need the let-down brought on by a quiet vacation; others seek the same pitch of excitement as exists in their work-a-day world" (Lundberg, 1971, p. 75). However, nowadays wellness and well-being are key concepts since people are looking for better and healthier lifestyles. This new way of living is also affecting their choices regarding travel. Thus, when looking and choosing a new destination, people are starting to consider tourism products that have the potential to increase their well-being and that will help them feel good.

The world is facing a shift, and tourism should be getting a "makeover" with new and exciting challenges. While the world is changing, communities and individuals are growing and new types of tourists with new ways of being are emerging. Tourists are looking for unique experiences, for quality, for self-improvement, and for authenticity. Hence, it is vital to not only understand these changes, and study them, but also to find new ways to attract people to destinations and give them a memorable experience. This is only possible if they feel well in the place they chose to visit and maybe revisit again.

Acknowledgments—The initial research for this chapter was financially supported by ARDITI (Regional Agency for Development of Research, Technology and Innovation of Madeira, Project M1420-09-5369-FSE-000001, Madeira 14–20 Program) through a postdoctoral research grant of the first author. This paper is financed by National Funds provided by FCT (Foundation for Science and Technology through project UIDB/04020/2020). The authors would also like to thank the City Hall of Funchal for the authorization conceded to data collection and the group of volunteers who collaborated with the project team.

Chapter 7

SENIORS BOOK TRAVEL ONLINE
A Profile Study

Ana Isabel Moniz
University of the Azores, Portugal

Teresa Medeiros
University of the Azores, Portugal

Osvaldo Silva
University of the Azores, Portugal

José Mendes
INTELECTO – Psicologia & Investigação, Portugal

Abstract: Senior travelers are an attractive market segment and increasingly participate in an internet-mediated society. This chapter examines the profile of senior tourists who booked their trip to the Azores using online travel agencies. The purpose is to analyze their motivation factors based on the travel motivation scale for senior tourists, using 17 items related to the motivations for visiting the destination, and to ascertain whether there are different groups of senior tourists based on sociodemographic characteristics, travel motivations, and experiences performed. Using a structured questionnaire and a sample of senior tourists, three distinct clusters are obtained. **Keywords**: Senior tourism; trip booking; online travel agencies; motivation; senior travelers; experiences in the destination

Managing Destinations: From Theories to Practices
Bridging Tourism Theory and Practice, Volume 14, 85–96

Published under exclusive licence by Emerald Publishing Limited
ISSN: 2042-1443/doi:10.1108/S2042-144320240000014007

INTRODUCTION

Studies suggest that longevity is on increase and a new paradigm in communication and interpersonal relationships is emerging, with repercussions on the way of life of people, including the elderly. Senior tourism is expected to grow by 18% by the year 2050, which leads organizers of tourism business to develop elderly-oriented products, such as cultural, emotional, social, and well-being activities (Santos & Bertoldi, 2012). Allied to this growth is the technological progress of Information and Communication Technologies (ICTs), which is having a strong influence on the overall growth of tourism, enabling consumers to identify, personalize, and purchase tourism products (Bethapudi, 2013; Morais, Cunha, & Gomes, 2013; World Tourism Organization & European Travel Commission, 2011). One of the problems related to the use of ICTs on the part of the senior tourist is due to the fact of the rapid evolution of these technologies (Lee & Groves, 2014) and the discontinuity between consumer learning models (traditional versus digital) and the new languages of being in the world. However, Mendes (2019) argues that the adaptation of ICTs to the elderly allows them to participate actively in social movements and follow technological developments.

Medeiros et al. (2020) and Santos and Bertoldi (2012) indicate that the growth of the senior tourism market is due to the fact that the elderly have more free time to visit new places, establish new social relationships, enjoy tourism products in longer stays, and feel the desire and need for unique, pleasant, and rewarding experiences. Although "new" senior tourists are more technologically savvy compared to previous generations (Klimova, 2017), access to information through tourism-oriented ICTs should become more dynamic and valid for any type of audience and support information technology (Puiggròs, Tort González, & Fondevila, 2017). Ruiz-Gómez, Navío-Marco, and Rodríguez-Hevía (2018) argue that the factors that make the use of ICTs in e-tourism probable vary according to economic development, with a degree of divergence over its use at European level.

Losada, Alén, Domínguez, and Nicolau (2016) refer that, in addition to the increase in the number of senior tourists, there is a change from more traditional vacations (sun and beach destinations, cruises) to holidays more based on experiences. These authors further argue that although the demographic profile of the senior tourist is important for the development of new products (demanding improved accessibility, infrastructure, and tourist services), these limitations are not decisive. The senior tourist is considered extremely heterogeneous, presenting different motivations for travel and with different lifestyles (Pesonen, Komppula, & Riihinen, 2015). The motivations of the senior tourist focus more on health (Tooman, 2015) and well-being. Senior trips are motivated mainly by visiting friends and family and looking for a pleasant climate due to the positive effect it has on the lives of the elderly (Alén, Nicolau, Losada, & Domínguez, 2014).

Senior tourism is noted in national and international flows (Cavaco, 2009; Martínez-Garcia, 2013; Santos & Bertoldi, 2012). Its growth has contributed to a greater understanding of the psychological aspects of elderly tourists, and it is also found that both positive and negative affective states influence their motivation (Jang, Bai, Hu, & Wu, 2009; Medeiros et al., 2019, 2020). According to Vigolo, Siemoni, Cassia, and Ugolini (2018), their satisfaction with the destination comes from motivations related to the search for pleasure, relaxation, physical activity, and culture. Recent studies on senior tourism in the Azores show that domestic senior tourists present self-knowledge and well-being as the main motivation, and those younger tourists (55–64 years old) attach greater importance to knowledge about the destination, personal learning, and sociability and fun (Medeiros et al., 2019; Moniz, Medeiros, Silva, & Ferreira, 2020; Silva et al., 2020).

The "modern" tourist (online) is more demanding and seeks direct interaction with tour operators, in order to satisfy their needs and expectations (Oneto, Ferreira, Giovannini, & Silva, 2015). Graziano and Privitera (2020) argue that augmented reality (AR) is perceived by tourists as a facilitator of new knowledge and promotes rapid interactions with cultural heritage, offering several opportunities to discover cultural heritage. This chapter assesses the use of ICTs by senior tourists, analyzing the characteristics, motivation, life satisfaction, and experience of those who booked the trip through online travel agencies. It employed a quantitative approach using a questionnaire to senior tourists, aged 55 years or over, who visited the Azores in the last 2 years. The objectives are to understand the main characteristics of these tourists, and tourist's degree of satisfaction with life and travel experience; to identify distinct segments of senior tourists (clusters), characterizing and defining their profile, according to their sociodemographic variables, their motivations and their experience; and to assess the differences between the segments (clusters) visiting the Azores.

ICTS AS ESSENTIAL TOOLS FOR SENIOR TOURISTS

Since the 1950s, tourism has been emphasized at a global, national, and regional level. In this significant existent touristic offer, tour operators have had to adapt their services and promote their products to satisfy tourists. ICTs' adaptation to the tourism activity is inevitable, thus emerging e-tourism (Serra, 2008). E-tourism is characterized as the digitization process of the tourism industry that promotes proactive and innovative management, as well as integrates and develops new technologies in tourist organizations (Kazandzhieva & Santana, 2019). The innovation of ICTs designed and optimized for tourism facilitates and makes routes more attractive, promoting cultural heritage, nature, and possible places of interest to the tourist

(Pierdicca, Paolanti, & Frontoni, 2019), also contributing to integrate sustainability concepts in tourism (Boavida-Portugal, Rocha, & Ferreira, 2016).

ICTs have been shown to be essential tools for senior tourists, as they positively influence travel expectations (Ramos-Soler, Martínez-Sala, & Campillo-Alhama, 2019). In fact, they are increasingly demanding, and there is still a need to customize offerings that match their expectations and understand what their choices are in the future (Patterson & Balderas, 2020). Tourism in the Autonomous Region of the Azores has been expanding in recent years. Senior tourism market has accompanied this growing trend, having represented 25.89% of the total between the years 2010 and 2018 (Azores Tourism Observatory, 2020). This evolving market sector is associated with the stakeholders' awareness and ability to (re)create existing resources and to personalize (adaptive capability engagement) products and services (Kelly, Kelliher, Power, & Lynch, 2020). For example, a study carried out by Kim and Preis (2016) found that senior tourists considered ICTs to be effective while traveling, allowing them to have fun, to enjoy new experiences, and to improve their self-satisfaction. In fact, they seem to be increasingly prepared to use the internet for tourism purposes, showing greater confidence, hedonic motivation, and performance expectations that, in its turn, influence their attitude toward these technologies' usage (Moura, Gosling, Christino, & Macedo, 2017). In addition to the fact that senior tourists have a medium level of knowledge in accessing ICTs, it appears that its use has an important effect on their motivations and intentions to travel (Wang, Wu, Luo, & Lu, 2017).

An analysis of the perspective of stakeholders on seniors visiting the Azores reveals that inclusive tourism is considered as extremely important (Medeiros et al., 2021a). For example, Cranmer, Tom Dieck, and Fountoulaki (2020) argue that innovative technology (AR) can integrate into existing marketing materials, improving information for tourists and boosting sales of products and services. ICTs can play an important role in accessible tourism (Eusébio, Teixeira, & Carneiro, 2021), allowing seniors to visit certain places on the islands of the Azores with limited accessibility.

ICTs have proven to be important in tourism, allowing tourists to maximize the length of stay, combining shopping and leisure activities for a global tourism experience. This sort of technologies may be desired by their users since it allows saving time and presents information in real time about possible constraints, such as traffic or high number of visitors (Huettermann, Thimm, Hannich, & Bild, 2019). ICTs' integration in tourism enables stakeholders to meet senior tourists' expectations, positively influencing their final perception about the experiences they had during the trip (Ramos-Soler et al., 2019). Medeiros et al. (2021a) report that one of the changes most recognized by stakeholders is due to the promotion and commercialization of the destination.

Mendes, Boavida-Portugal and Moniz (2023) carried out a study that identified a concern about the use of ICT by tourist activity, to implement strategies that analyze the needs of tourists and to create hedonic experiences. A study conducted by Shao, Bai, Shu, and Joppe (2020) reveals that innovative technologies enable stakeholders to better manage planning and allow tourists to have deeper experiences. Senior tourists' acceptance toward this type of technology is influenced by factors such as perceived ease of use, perceived usefulness, perceived cost, perceived enjoyment, and perceived immersion. Ramos-Soler et al. (2019) advance that these tools' development and configuration must be adapted to senior tourists' profile and must guarantee a positive sustainability of destinations.

Recently, a study with a sample of about 1,000 senior tourists who visited the Azores concluded that the most positive aspects associated with their trips are characterized by contact with nature/climate, culture, heritage, gastronomy, safety, accessibility, information, and activities adapted (Medeiros et al., 2021b). Considering that senior tourists are recognized by the W generation (Ramos-Soler et al., 2019), it is important to use ICTs in order to create mobile applications that provide information about the destination and suggest possible good routes for senior tourists visiting the Azores.

Study Methods

The quantitative and exploratory study is based on a sample (n = 252) of senior tourists (≥55 years) who booked the trip with an online travel agency when visiting the Autonomous Region of the Azores. For data collection, a structured questionnaire was used to characterize senior tourists, in which the travel motivation scale for senior tourists was applied, considering it allows to measure three dimensions (knowledge and personal enrichment; sociability and fun; and identity exploration and search for well-being) of motivations for choosing the Azores as a destination (Silva et al., 2018).

The questionnaire includes four groups of questions related with sociodemographic characterization (sex, age, nationality, educational level, living with partner, retirement situation, household composition); related with the trip (schedule, who travels, island (s) visited, choice of accommodation type, length of stay, among others), regarding satisfaction with life; and regarding satisfaction with the visit. The motivation scale consists of a set of 17 items regarding the motivations for visiting the Azores. Each of the items referring to the motivations for the visit was evaluated on a 5-point Likert scale (1 = not important, 2 = slightly important, 3 = indifferent, 4 = important, and 5 = very important). Satisfaction with life and with the visit were both assessed on a 5-point Likert scale (1 = not at all satisfied, 2 = slightly satisfied, 3 = moderately satisfied, 4 = very satisfied, and 5 = completely satisfied). Ethical procedures were carried out to guarantee confidentiality.

Data collection was conducted between January 2017 and July 2019, using cluster sampling according to geographical area of residence, namely the islands of the Azores, Portugal mainland, and the United States of America. The data were processed using IBM SPSS Statistics (Version 25), using several statistical methods, namely the multiple correspondence analysis model, the nonhierarchical classification analysis (k-means method), and the Kruskal–Wallis test (nonparametric test). The model is a multivariate technique that makes it possible to investigate the existence of an association between more than two categorical variables. The components resulting from it are used, making the respective articulation with the nonhierarchical classification analysis, using the k-means procedure, which will help to know in more detail the groups (profiles) that can be found based on the dimensions considered in the model.

Study Results

The sample includes tourists of both sexes, mostly male (50.4%), aged between 55 and 88 years old, and particularly in the age group of 55–64 years old (63%), with higher education (44.5%), in retirement situation (53.1%), living with a partner (80%), their household is composed of 1–3 elements (71.8%), and the majority are foreigners (66.3%) from Europe (66%) (Figure 1).

In this study, the participating senior tourists reveal a high level of satisfaction with life (69.1%) and demonstrate great satisfaction with the visit to

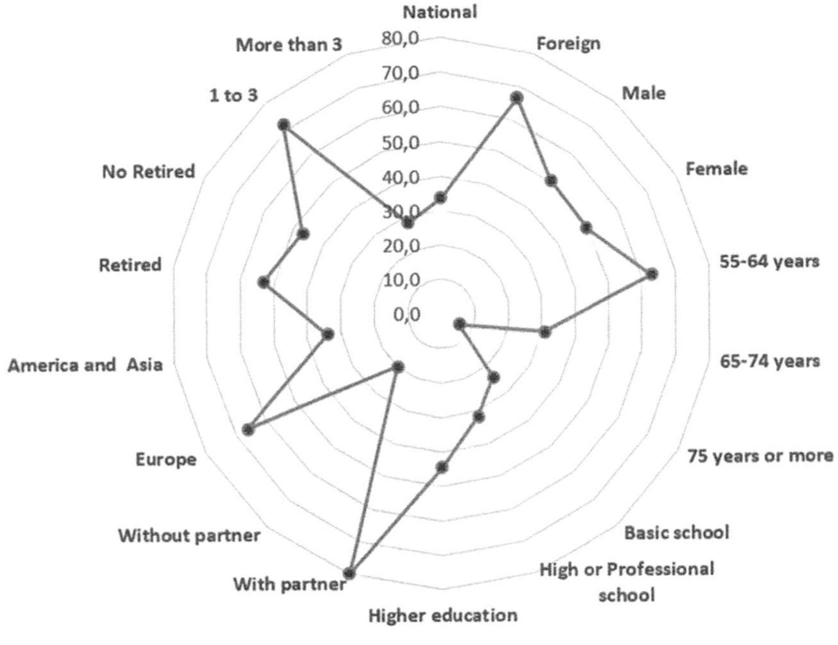

Figure 1. **Sociodemographic Characterization of the Respondents.**

the Azores (61.8%) (Figure 2). Further, 84.6% prefer to travel with family/friends, and 61.8% are very satisfied with their trip.

Using the multiple correspondence analysis and articulated with a nonhierarchical model (k-means), three distinct groups of senior tourists are characterized (Figure 3). The first group (C1 = 21.8%), "nostalgic contemplatives," is formed mainly by female domestic tourists, with lower education and greater health problems. In the second group (C2 = 33.3%), senior tourists are mostly foreigners, aged between 55 and 64 years old and have secondary education or vocational training as their academic qualifications. They are labeled "adventure seekers," as they value adventure and exploration activities.

The largest group (C3 = 44.9%) corresponds to the "experienced in traveling." They are mainly characterized by being mostly foreigners and male tourists, aged between 55 and 74, being the group with the highest academic qualifications (higher education). A study by Losada et al. (2016) states that variables, such as sex and age, do not have a great influence on the activities and motivations of senior tourists; however, the educational level, profession/occupation and economic status have a great influence in the behavior and travel frequency so they can be considered determinants of the lifestyle of this cluster. The decision to participate in tourism is affected by the level of education, and such knowledge is useful for defining appropriate strategies of market segmentation demand. Tourists with high level of education access information more easily and have greater motivation and desire to discover new destinations and enjoy new experiences (Bernini & Cracolici, 2015; Medeiros et al., 2019, 2020; Silva et al., 2021).

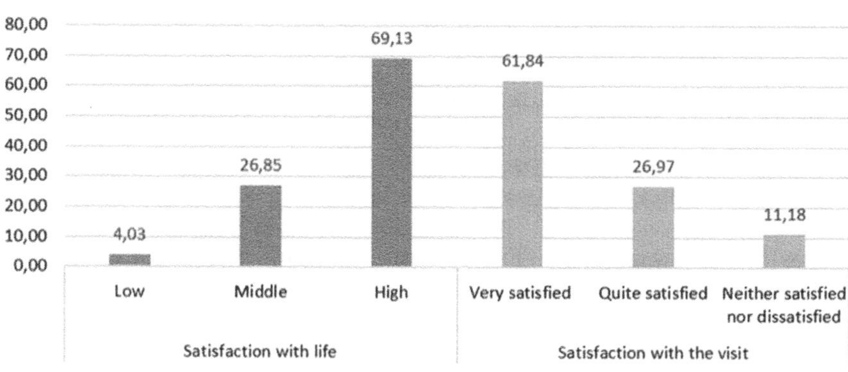

Figure 2. Satisfaction With Life and Satisfaction With the Visit.

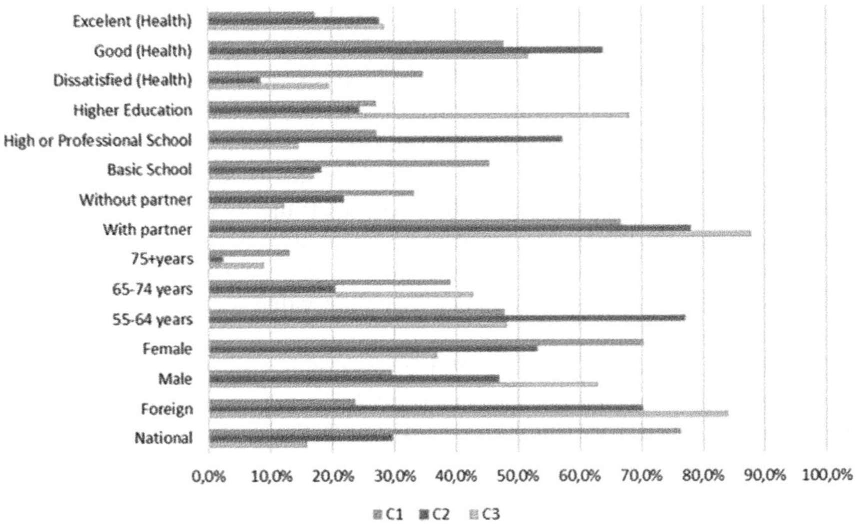

Figure 3. Sociodemographic Characterization of the Clusters.

Regarding motivations, the group "nostalgic contemplatives" (C1) is characterized by giving higher importance to both factors "knowledge about destination and personal learning" (62.2%) and "Sociability and fun" (97.6%). In terms of travel experiences, and when compared with the others, this cluster stands out for preferring contemplation of nature (92.9%), thermal baths (60%), and traditional festivities. In order to increase attractiveness, they suggest improved accessibility to the destination. The "adventure seekers" group (C2) attaches lesser importance to the "knowledge about destination and personal learning" factor (82.5%) and, in contrast, greater value to "sociability and fun" (84.4%). It is the group that most enjoyed the nature (96.6%), followed by thermal baths (91.9%) and heritage (80.4%). In terms of motivational factors, the third cluster, the "experienced in traveling" group (C3), consider nonimportant. They are the most pleased with the visit and with most previous visits to the Azores (51% visited more than 3 times). Preferences are very diverse, from learning and well-being activities to contact with the population. They recommend more investment in tourism education and training and improving the quality of services. As for the activities carried out, 83.3% of the respondents reveal high satisfaction in activities related to contact with nature (Figure 4).

It is possible to verify that this group (C3) is more satisfied now than it was on previous visits to the Azores (51% of the surveyed tourists had visited the Azores more than 3 times). These findings may be related to the fact that familiarity, learnings, and experiences with previous visits have an indirect influence on online shopping, as it allows tourists to dispel uncertainties when choosing the destination and purchasing the trip (Ruiz-Gómez et al., 2018).

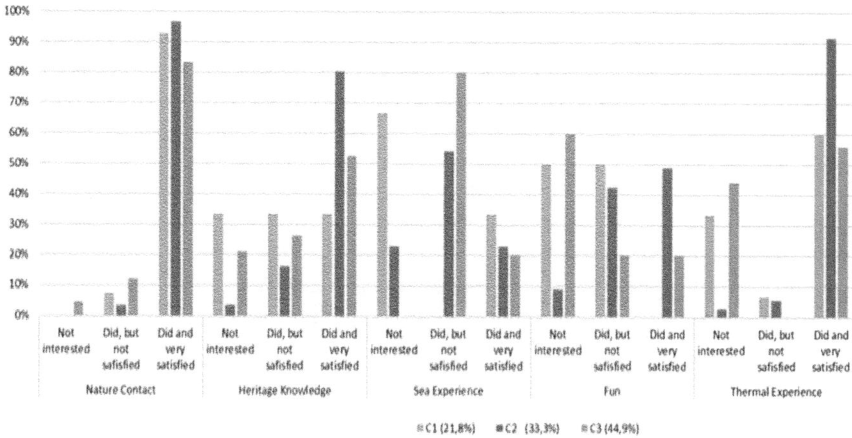

Figure 4. Experiences in the Destination by Clusters (C1, C2 and C3).

This satisfaction can also be related to hospitality (Crotti & Moretti, 2019; Medeiros et al., 2020) since in the opinion of Ip, Leung, and Law (2011), ICTs can be an asset in the relationship between tourists and tourism professionals. However, it appears that the use of ICTs in tourism seems to gain ground in the field of research (Law, Leung, & Chan, 2019), as they promote the marketing and management of destinations (Mandić & Garbin Praničević, 2019). The reasons, attitudes, destination selection criteria, and tourist activities in senior tourism do not change dramatically during the stages inherent in the aging process (Chen & Shoemaker, 2014; Tooman, 2015). For Klimova (2017), the use of ICTs reflects greater confidence and greater independence for the senior tourist during the trip.

By applying the Kruskal–Wallis test, it was found that there are statistically significant differences between the three clusters of senior tourists, according to their motivating factors. In all cases, and based on the mean values, the "nostalgic contemplatives" give the most importance to these three factors, and the "experienced travelers" are which comparatively attaches the least importance to these factors (Table 1). These results are in line with the study by Vigolo et al. (2018) stating that the motivations vary markedly, depending on the age of the senior tourists.

According to Tooman (2015), various approaches are known about the profiles, needs, and requirements of senior travelers (safety at destination, the reputation of tour operators, safety and health of seniors, the development of accessibility easy and convenient, promotions and accommodation packages

Table 1. Kruskal–Wallis Test.

Motivations	*n*	C1		C2		C3	
		Mean Values	Kruskal–Wallis Test	Mean Values	Kruskal–Wallis Test	Mean Values	Kruskal–Wallis Test
KDPL	196	145.54		95.69		73.14	
SF	176	130.82	41.24*	98.58	61.46*	51.25	16.12*
WbSk	106	69.20		59.30		39.71	

Note: KDPL = Knowledge about destination and personal learning; SF = Sociability and Fun; WbSk = Well-being and self-knowledge. $^{*}p < 0.01$.

that are exciting, creative, and value-oriented, personalized service, friendliness, support, and hospitality, socializing with others, healthy food, and activities, among others.). For this reason, it is important to create beneficial, simple, and compatible digital platforms with senior tourists (Kim, Lee, & Contractor, 2019). Networking is a fundamental policy for optimizing senior tourism in the Azores (Medeiros et al., 2021). ICTs incorporated in tourist services can contribute to improving the management of tourists and their experiences organizers of tourism business (Hughes & Moscardo, 2019), capable of transmitting information on a large scale, providing tourists with updated content and providing useful information on their needs, habits, and preferences (Pierdicca et al., 2019).

Medeiros et al. (2021a) mention that one of the concerns presented by stakeholders is focused on improving access to certain products and tourism attractions. In fact, the marketing/design of ICTs for senior tourists needs greater investment so that these technologies can meet their needs (Pesonen et al., 2015). Mihajlović and Strugar (2015) mention that most tourists look for personalized services that allow, for example, easy access to information and a wide selection of services and products, increase the quality of the experience, among others.

Rethinking the use of ICTs in senior tourism, in addition to contributing to a greater involvement of seniors with companies, can also provide them with a new experience when deciding which places to visit. Puiggròs et al. (2017) argue that ICTs should make tourists more active, articulating mobile devices with them, allowing a greater level of interactivity.

CONCLUSION

In the last two decades, a transformation has been witnessed because of the ICTs. ICTs (virtual reality, AR, or mixed reality) bring new paradigms for tourist activity (Egger & Neuburger, 2020) in which these technologies have evolved both in immersion and in the sensation of telepresence, creating valuable opportunities for the innovation of tourism activity and promoting new experiences for tourists unavailable in the real world (Loureiro, Guerreiro, & Ali, 2020). Medeiros and Mendes (2021) mention that AR, despite being increasingly present in the daily lives of tourists, seems to be the target of little research on how this technology is used by tourists.

The development of a wide range of new tools and services has facilitated global interaction between players around the world (Morais et al., 2013). More and more senior tourists are planning their trip individually, as a family or as a small group of friends, using ICTs to book their ticket and to plan the places to visit by making their own program. Despite acknowledgment by several studies analyzing senior tourists, there is a lack of empirical evidence on the profiles of tourists that use online travel agency for booking.

This research segments and characterizes senior tourists who used ICTs to book their trips to the Azores Region in the past 2 years using an online travel agency, according to travel motivation factors. The participants are mostly foreigners, coming from Europe, with a high level of education, aged 55–64, retired, and show high satisfaction with life. Regarding their motivations to travel to the Azores, three clusters were found: (C1) the nostalgic contemplatives group, that highlights knowledge about destination and personal learning and sociability and fun, preferring contemplation of nature, thermal baths, and traditional festivities; (C2) the adventure seekers group that give higher importance to sociability and fun, enjoy the nature, thermal baths, and heritage; and (C3) the *experienced in traveling* group that visited more than three times the destination. This last cluster does not give importance to the motivation factors preferred by the other groups, favoring a diverse set of activities such as learning and well-being activities, contact with the population, and valuing contact with nature.

Considering the increased need for destination management organizations to manage and monitor the dynamics of tourism in destinations, it is of foremost importance to broaden the knowledge about these types of tourists, the attributes that motivate them and shape their travel behavior, and their satisfaction with the experience. The results of this study will contribute to the design of adequate marketing and management strategies. Considering the increase in the average life expectancy, the fact that future seniors will be progressively more able to use ICT and the influence that these technologies can have on motivation and satisfaction in making a trip, it is relevant to implement innovative technologies (AR) in senior tourism. At the same time, stakeholders must prepare for an increasingly technological tourism market, giving their contribution in developing smarter destinations.

Acknowledgments—This study was carried out within the framework of the University of the Azores research project "TuSénior 55+ - Senior Tourism and Wellness in the Azores Destination: Creating a Cultural Product," funded by the PO2020 Azores program and the Azores Regional Government, with reference ACORES-01-0145-FEDER-00002 Senior 55+.

DESTINATION STAKEHOLDERS

Chapter 8

TOURISM DEVELOPMENT Al-Juhfa, Saudi Arabia

Ali AlGassim
Umm Alqura University, Saudi Arabia

Akhmad Saufi
University of Mataram, Indonesia

Diswandi
University of Mataram, Indonesia

Noel Scott
Edith Cowan University, Australia

Abstract: Involving residents in planning for sustainable tourism can help gain their support. This chapter examines residents' support for tourism development in Al-Juhfa Village, Saudi Arabia. It identifies three ways to gain residents' support. The first is inclusion of residents' heritage and agricultural activities in tourism planning, the second is development of infrastructure and tourism facilities that benefit locals, and third the empowerment of destination residents. Inclusion of these in the masterplan will strengthen resident support for tourism development. Recommendations for policymakers and destination managers on developing local attractions and strategies for empowering resident are provided. **Keywords**: Tourism development; heritage tourism; community empowerment; Saudi Arabia; sustainable tourism; resident perception

Managing Destinations: From Theories to Practices
Bridging Tourism Theory and Practice, Volume 14, 99–110

Published under exclusive licence by Emerald Publishing Limited
ISSN: 2042-1443/doi:10.1108/S2042-144320240000014008

INTRODUCTION

The tourism industry is used by many countries to create jobs and improve economic development (Wall & Mathieson, 2006). Tourism development however requires careful planning as it can cause both positive and negative impacts on the economic, sociocultural, and natural environment (Weaver & Lawton, 2014). Any development can have positive effects on the local communities but also negative sociocultural impacts (Mbaiwa, 2005). Tourism activities may create overcrowding, traffic, and a higher cost of living (Brunt & Courtney, 1999), and stimulate the emergence of gambling, begging, and abuse of drugs, thereby threatening the culture and traditions of local communities (Lin, Chen, & Filieri, 2017). The development process and its concomitant acculturation can threaten cultures and traditions by weakening local values and cultures (Zhuang, Yao, & Li, 2019). Tourism may bring changes to many aspects of local life, such as cultural identities, family relationships, traditional lifestyles, and moral values (Zhuang et al., 2019). Furthermore, tourism can create threats to safety and security and lead to environmental damage (Jackson & Inbakaran, 2006).

Sustainable tourism aims to decrease the negative impacts of tourism activities in order to achieve a balance between economic viability, ecological sustainability, and ethical and socially equitable development (Pan et al., 2018; UNWTO, 2005). Sustainable tourism must satisfy the needs of destinations and tourists on the one hand, and create opportunities for more future development and maintain the integrity of both the destination's cultural heritage and environment and biodiversity on the other hand (UNWTO, 2013). Development of tourism must improve the satisfaction of tourists and destination sustainability (Sharpley, 2000).

Tourism can contribute to a destination's economy at the micro and macro level. It can provide direct or indirect employment for destination residents (Cukier, 2002). Local people can find jobs in such tourism sectors as in hotels, restaurants, and travel agencies. Tourism also stimulates creation of indirect jobs such as in medical, retail, and related services (Cukier, 2002). Its economic impacts enhance local people's standard of living and stimulate growth of the private sector which, in turn, creates contributions to government income through tax and an increase in gross domestic product (GDP) (Sharpley & Telfer, 2002).

Tourism also has positive and negative impacts on the social well-being of destination communities. It brings additional investment and spending, which can play a significant role in supporting the creation of facilities and infrastructure, consequently enhancing life quality for residents. Transport, roads, water, energy, health services, leisure, and entertainment, as well as outdoor facilities, shops, garages are among such services and facilities (UNWTO, 2013).

Moreover, many elements contribute to achieving a sustainable society (UNWTO, 2005). First, the quality of life in local communities can be improved. This element includes access to resources, social structures, facilities and services, and health systems. At the same time, exploitation and degradation, whether social or environmental, can be prevented, or at least reduced. Second, the economic and social benefits that are gained from tourism can be distributed fairly to the local community. These benefits can provide the poor with better opportunities, services, and income as they can find work or open small tourism enterprises (Saufi, Reid, & Patiar, 2017). Third, tourism can increase job opportunities for locals, whether this relates to the number of available jobs or to the quality of these jobs. This includes increasing pay level, improving the conditions of service, and widening the availability of services to all members of the local community without any sort of discrimination based on race, gender, or ability. Finally, tourist fulfillment and satisfaction are achieved when everyone is provided with a safe experience that satisfies their needs.

Providing job opportunities for the local residents in any destination is one of the main ways in which tourism can contribute to improving the quality of life in the host community (Mihalič, 2002). While the industry makes a significant contribution to the economy of many countries, jobs are usually low-paid and in some cases have poor working conditions (Weaver & Lawton, 2014). This is because work in the industry is often characterized by seasonal employment and part-time work (Cukier, 2002; Hampton, 2005) and a very high employee turnover rate (Scheyvens, 2003; Weaver & Lawton, 2014).

In many countries, important cultural and historic sites appeal to tourists. The management and conservation of these sites often relies on the income produced from tourists to them. They can damage the fabric of many sites, which, in turn, negatively impacts the quality of the tourist experience (Hampton, 1998, 2005). Hence, it is important to raise the awareness of host communities and teach them how to value the distinctiveness of their destinations, their native cultures, authentic historical heritage, and traditions (Hampton, 2005; Scheyvens, 2003).

One of the principal aspects of sustainable development lies in respecting and appreciating cultural diversity (UNWTO, 2005). In this context, tourism can play a vital role in preserving and maintaining cultural and historic heritage (Hampton, 1998, 2005). It can also enhance all sorts of creative activities in communities, including arts and crafts (Thompson, 2004). Thus, tourism can be a means to generate a source of income from local culture. In this manner, it can urge communities to have a high regard for their heritage (Thompson, 2004). Nevertheless, measures are needed to prevent practices that falsify and degrade heritage and culture (UNWTO, 2005, 2013).

A sustainable environment relates to the material structure of tourist sites as well as to their appearance and aesthetic quality. From an ecological point of view, this dimension is an end in itself. It also has a positive effect upon the well-being of the local community (UNWTO, 2013). Moreover, it contributes to the success of the industry in the long term. This is due to the fact that when destinations are physically attractive, they appeal to and attract more tourists (UNWTO, 2013).

Therefore, a sustainable development plan seeks to preserve and boost the quality of urban and rural landscapes and guard against degradation of the visual and physical aspects of destination environments (Gössling, 2002; Gössling, Hansson, Hörstmeier, & Saggel, 2002). One way to maintain landscape quality and prevent the degradation of destinations is to implement policies that regulate the nature and location of new development projects (Gössling, 2002). Such policies should also focus on constructing facilities related to tourism, including accommodation and infrastructure (UNWTO, 2005, 2013). This chapter examines planning for tourism development in Al-Juhfa Village, Saudi Arabia.

Context of the Study

The Kingdom of Saudi Arabia has unique resources, and the tourism and hospitality industry is considered as one of its emerging economic sectors (AlGassim & Abdulaziz, 2021; AlGassim & Abuelhassan, 2021). The Kingdom has recently begun to focus attention on tourism. The country is the cradle of Islam, and millions of Muslims visit the Holy Mosque in Makkah to perform the Hajj and Umrah, as well as visit Prophet Muhammad's Mosque in Al-Madinah. The Kingdom's economy depends on oil, which represents more than 85% of its GDP, leading to the marginalization of tourism development (AlGassim, Barry, & McPhail, 2012). The decline in demand for oil and fluctuations in its price have emphasised the importance of economic diversification into the tourism industry. The government has developed several strategic initiatives and projects to enhance the Kingdom as a destination for Muslims.

Saudi Arabia is considered among the leading destinations in the Middle East, although tourism is traditionally limited to Hajj and Umrah activities. According to the UN World Tourism Organization, in 2018, the number of international arrivals reached 15.3 million (UNWTO, 2019), with Hajj and Umrah providing 45% of this total. Domestic tourism has also grown, with a compound annual growth rate of 14% since 2010 (UNWTO, 2019). The number of domestic flights grew to 44 million in 2017 (SAGIA, 2019). Inbound tourists spent US$25 billion (SR 94 billion) in 2016, while domestic tourists spent US$ 14.3 billion (SR 53.6 billion) in the same year (SAGIA, 2019). The country is currently ranked among the top 20 in the world in terms of tourism's total contribution to the GDP (SAGIA, 2019).

Saudi Arabia has recognized the prominent role that tourism can play as a driving force for economic and human development in its 2030 Vision. One of its goals is to transform tourism and hospitality into a leading economic industry. The number of Hajj and Umrah pilgrims is targeted to grow from the current 8 million to 30 million by 2030. The government has also changed the name of the Saudi Commission for Tourism and National Heritage to the Ministry of Tourism, established the Saudi Commission for Tourism, begun issuing tourist visas, and encouraged tour programs after Umrah.

For the Kingdom to increase the number of tourists, it is necessary to develop religious, archaeological, and Islamic heritage sites, increase the capacity of tourism projects, hotels, furnished apartments, and residential buildings, and increase tourism services. For this reason, global and local hotel companies are increasing their investments throughout the country, especially in Makkah and Al-Madinah. Yet, the prospects for development of tourism destinations outside of the two cities remain relatively unknown. This study investigates local residents' preferences regarding tourism development in Al-Juhfa Village. The findings will provide the Saudi government with information on which to develop sustainable tourism in this historical village and miqat place near Makkah.

TOURISM POTENTIAL OF AL-JUHFA VILLAGE

The village of Al-Juhfa is in the west of Arabia, between Makkah and Al-Madinah, to the north of the former, and to the west-south of the latter. It is 13 kilometers away from the sea on the main road of Hijaz Tihamah, on the coastal plain between the Red Sea and Taif Mountains (Abu-Al-Fida, 1850). Al-Juhfa Village is within Rabigh Governorate, about 177 kilometres away from Makkah and about 20 kilometers from Rabigh, the capital of the Governorate (Al-Baladi, 2010).

Al-Juhfa's small population of about 110 live in 36 houses, each consisting of 1 floor and from 2 to 4 rooms. The rest of the village's residents have moved to work or study in Rabigh, Jeddah, or Makkah. The average number in a family is four. The residents are from the Harb tribe, one of the largest tribes that settled in the Arabian Peninsula in the Kingdom of Najd and Hijaz more than 200 years ago. All these families are descendants of nomadic Bedouins, who used to move from one place to another, especially in summer when the area became devoid of natural resources due to a lack of rain.

Al-Juhfa is characterized with an extremely high maximum temperature exceeding 40 degrees Celsius at noon in summer. The temperature is lower after sunset at about 28 degrees Celsius. On the other hand, the temperature in winter is moderate, with a minimum of less than 16 degrees Celsius. The Al-Juhfa area has high humidity in late summer and early autumn, sometimes up to 90% because of the high evaporation rate of the Red Sea. Peeling and splitting rocks litter the area of Al-Juhfa due to the variations in temperature and humidity between day and night, and between summer and winter.

Al-Juhfa was an important location in ancient times, as it was an important station between Makkah and Yathrib (Al-Madinah). It was also a station for Hajjis (pilgrims) heading for the Holy Kaaba by land from the north-west. The founding of the village dates to around 2500 BCE. The old city extends to the west of Al-Juhfa Fort, and there are numerous archeological remains such as camel resting places. The village was a well-organized and spacious urban area (Bakr, 1981), with many architectural features such as large houses, a large water pond, a water basin, many wells, a well-built fort, two gates, and a market (Khardadiyah, 1889).

Al-Juhfa Village flourished because it was a miqat for Hajjis on the coastal path. A miqat place is at the boundary around Makkah where pilgrims must enter the state of iḥrām, a state of consecration in which certain permitted activities are made prohibited. Al-Juhfa Mosque was used as a miqat for countries such as Morocco, Tunisia, Al Jazair, Egypt, and the Levantines (Bakr, 1981). The importance of this miqat lies in the three holy mosques in this region. Historical sources state that there was a mosque in Al-Juhfa called the Mosque of Prophet Muhammad (PBUH), also called the Azura Mosque (Al-Baladi, 2010). A second mosque, the Mosque of Imams, lies at a distance of three or four miles from Al-Juhfa Village (Al-Baladi, 2010), at the edge of a stream, but its exact location is unknown. A third mosque was known as Ghadir Khumm Mosque and located in the middle of a rivulet.

Al-Juhfa has access to large amounts of water for agriculture and drinking, making it a destination for those who came from distant and dry areas (Balsoud, 2009). The village had an outstanding commercial position because it was the intersection of the caravan routes between Makkah and Al-Madinah, between a seaport and the two holy cities, and on the caravan routes coming from and going to the Levant, Egypt, and North Africa. It has a strategic location between Makkah to Al-Madinah for Hajjis and those undertaking Umrah. These characteristics suggest that the village can be developed for tourists and so become an economic resource for the Kingdom of Saudi Arabia in general and for the Rabigh region and the local population of Al-Juhfa Village in particular. Involvement of residents is an important aspect in the process of any sustainable development strategy.

Heritage and Agricultural Activities

Informants in the village identified two main attractions: the heritage of Al-Juhfah Village and the agricultural activities of the residents. The informants pointed out that the old, buried village within the historical site of Al-Juhfa must have contained a village market. That market could be rebuilt and equipped with products that tourists and local people would like to buy. An alternative option is to build a market in the village, using the same type of stones from which the village was built.

The agricultural activities of local residents of Al-Juhfa Village can provide unique and distinctive attractions for tourists. The people of the village stated that grazing and agriculture were, and still are, part of their daily life. Therefore, this legacy can be used to develop agriculture-based attractions, by organizing tours to their farms which are prolific in the rainy season, especially watermelon farms. These agricultural products can support the attraction by providing food and beverages for tourists. Residents can establish and operate small enterprises such as restaurants, gift and handicraft shops, and grocery stores. Residents can also provide accommodation (hostels) for tourists. These businesses could provide an income for the local population and provide them with job opportunities. Currently, residents do not have sufficient experience in establishing and managing their owned tourism enterprises and need help to initiate and develop these projects.

Infrastructure and Tourist Facilities

The village development plan aims to build appropriate physical infrastructure and tourism facilities within Al-Juhfa Village. This will include new transportation infrastructure, refurbishment of a Miqat Mosque, a heritage museum, theater building, thematic park, public services, accommodation facilities, and a business area.

The Al-Juhfa tourism development should be supported by provision of adequate transportation, which requires good road infrastructure. Nowadays, the village has neither public transportation nor access roads. For accessibility purposes, it is important to provide public transportation, such as buses and taxis, to meet the needs of tourists and Umrah pilgrims who will visit the miqat and the village. Adequate transportation and road facilities will support the smooth arrival of tourists and businesses at Al-Juhfa. In addition, providing these facilities will increase job opportunities for the local people.

The Miqat Mosque now is relatively small. To meet the needs of a larger number of people visiting the Miqat to do Hajj or Umrah, the existing mosque needs to be renovated and equipped with better facilities. The old Miqat used since the time of the Prophet Muhammad (PBUH), which is currently located inside the women's prayer area, can be renovated to promote the Miqat of Al-Juhfa as well as for use by the village. This mosque is also useful as a place of prayer for tourists visiting Al-Juhfa Village and the local community.

The main attraction at Al-Juhfa is the heritage site. This heritage site should be preserved as a museum. If the site becomes a museum with restricted access, this will help to preserve it and provide a place where various historical stories are enacted, and ancient Arab and Islamic heritage artifacts stored. The site of the old village is believed to contain monuments and artifacts of the life of the people who lived there. These artifacts can be curated and interpreted in the museum and provide a tourism attraction. The museum can also introduce Arabic culture and Islamic civilization to tourists and communicate stories from the history of the ancient nations that controlled the village. The biography of Prophet Muhammad could be displayed in the form of visual stories or material artifacts.

The museum could also have a theater attached to display the area's cultural heritage visually and virtually. The design and construction of the theater building should be consistent with the characteristics of the local architecture. A theater can be used to various purposes including cultural shows, meetings, exhibitions of the traditional Arabic arts, and other important events. The theater could provide dramas including the biography of the Prophet and the Prophet's Companions to allow tourists learn about the beauty of Islamic teachings. The development and operation of the theater can provide jobs particularly for local youth. However, further research is needed to validate the economic feasibility of a theater development.

A Qur'anic thematic park can provide a unique attraction for people visiting Al-Juhfa Village. People who visit this park can learn about the *Qur'an* by looking at events that are exhibited, increasing their knowledge of *Qur'an* and Islamic teachings.

Al-Juhfa Village would also need public service facilities, such as a petrol station, medical center, and communication facilities. Currently, the closest public services available are in Rabigh, which is about 20 kilometers away from Al-Juhfa. These public services are needed to serve the local community, tourism businesses, and incoming tourists. Likewise, to maintain security, it is necessary to establish a police station. Furthermore, as a destination, the village needs adequate communication facilities which, in turn, contribute to the welfare of the local residents of Al-Juhfa and its surrounding communities.

Accommodation is critical to the tourism industry. Hotels and restaurants will attract tourists to visit and stay in Al-Juhfa. Since the village is on the main road for the people traveling between Makkah and Madinah, the availability of accommodation will increase the possibility of people stopping. The construction of hotels and restaurants must consider the environmental issues as accommodation and other tourism facilities will increase the need for water in Al-Juhfa. Hotel construction should not create a shortage of water for the surrounding locations. The latest technologies such as converting seawater into fresh water using solar power can be considered, given the location of Al-Juhfa, which is near to the Red Sea.

There are few existing shops in Al-Juhfa. Some located around the Miqat Mosque sell equipment and clothes for Umrah. The construction of a shopping center will be useful in meeting the needs of tourists visiting Al-Juhfa. The development of a business area will create employment opportunities for the community, both as entrepreneurs and employees.

However, the residents will need training and empowerment which should be provided by the government to play an important role in the business area either as a local supplier or business owner.

Destination Resident's Empowerment

Training is one of the main pillars for empowering residents. Training of Al-Juhfa residents provided by the government is needed to help prepare them to be entrepreneurs or workers in tourism and hospitality and related businesses. The importance of training increases because there is a shortage of workers specialized in tourism. Moreover, the importance of training lies in building the required skills and developing teamwork characterized by discipline and self-confidence. An empowerment program for the residents can include training as a tour guide, for small-scale tourism business operators, and in starting a business.

As a heritage tourist destination, Al-Juhfa requires qualified tour guides who have sufficient knowledge about the village and skills in how to professionally guide tourists, Hajjis, and Umrah pilgrims. The empowerment program for residents wishing to be tour guides can be in the form of training workshops. Workshop training should cover the curriculum necessary for a tourist guide, such as knowledge of the history of the Kingdom of Saudi Arabia in general and the Islamic monuments that exist in the Arabian Peninsula in particular, and specifically at Al-Juhfa Village. In addition, there is also a need to develop the guide's communication skills.

Sustainability training by the government is needed to provide the local community and people and companies with a basic understanding of sustainable business practices. The aim is to provide training so residents can maximize the economic benefits of tourism industry at Al-Juhfa and minimize negative impacts on the local culture and environment.

Training in entrepreneurship and small business management needs to be provided by authorities, such as the Ministry of Tourism and/or the Ministry of Heritage. The types of businesses needed in Al-Juhfa are small businesses that can be established by people with little capital. With small business training, more people will take part in the tourism industry at Al-Juhfa. This training can be provided free of charge to the public by bringing in trainers from universities and tourism schools in Saudi Arabia.

Start-up training is needed to help residents initiate tourism enterprises. Intervention by government can be in the form of financial support, particularly at the set-up stage of the tourism business. The residents of the village and its surrounding communities indicated in their responses that they would like to invest in small businesses when the destination is developed. Governmental intervention is required to stimulate business development by, for example, providing capital assistance for those with limited capital, providing training in entrepreneurship and other business skills, and assisting in feasibility studies of the proposed businesses. The financial assistance can be in the form of low interest loans. To attract investors to open businesses or invest in Al-Juhfa, the government could provide incentives to investors, for example, in the form of various licensing concessions or tax breaks.

These empowerment programs together are necessary in order to help residents to participate in the tourism development within their area. A mix of soft skills training and financial support for the destination residents is needed. Such training programs can help strengthen the abilities of the local people and the institution that will support the project development, especially with reference to tourism entrepreneurial skills, developing an institution to manage tourism at Al-Juhfa, small scale business training, and building a sustainable business system. The empowerment of entrepreneurship can help stimulate the emergence of local tourism businesses from within the local residents, who can be involved in development.

Sustaining Tourism Development. Tourism development can have both positive and negative impacts on community life (Egresi & Kara, 2018). When a community is small, its members tend to oppose tourism because they see that this development will occur at the expense of the services and facilities available for them. The permanent residents of small places have the right to benefit from tourism in addition to allowing the fulfillment of the wishes of tourists (Capenerhurst, 1994). It is important that the residents of the village have these opportunities.

Tourism development in Al-Juhfa can bring positive economic, sociocultural, and environmental benefits to residents, and the Kingdom. For tourism to be sustainable, the destination and businesses should be viable and competitive to guarantee their prosperity and their delivery of benefits in the long run. Tourism can lead to the economic prosperity of the local community, and prosperous enterprises play a vital role.

Residents of Al-Juhfa have a range of attitudes toward tourism development that are influenced by their perceptions of the future economic, sociocultural, and environmental impacts. The residents support tourism development if they perceive positive benefits of activities within their area to them (Alrwajfah, Almeida-García, & Cortés-Macías, 2019; Egresi & Kara, 2018). Thus, the perceptions of residents of Al-Juhfa illustrate the complex relationship between perceived economic and sociocultural impacts and support for tourism development (Boley, McGehee, Perdue, & Long, 2014). This range of attitudes suggests that different interventions by the government and institutions are needed to increase community participation and create opportunities to benefit from tourism for the villagers (Saufi, O'Brien, & Wilkins, 2014).

Most residents of Al-Juhfa Village do not perceive tourism development as bringing negative economic impacts. They think that tourism will make

Al-Juhfa a new economic center, creating direct income for the government and the local community, business assistance, work, and opportunities. The arrival of employers and workers from outside to the village is not perceived as an economic threat to residents for two main reasons. First, they believe that the government will issue pro-local policies to protect residents from competition and economic loss. Second, the industry requires many employees, and local workers alone cannot meet this demand, as they need tourism training, knowledge, and skills.

These results are like other studies on community's perceptions toward tourism where local communities tend to view the impacts of its development as positive. For example, a study of local communities in Australia found that residents accept tourism as long as this development does not threaten them and their role in the community, but if residents feel threatened, they will oppose its development (Jackson & Inbakaran, 2006). Similarly, Gursoy, Jurowski, and Uysal (2002) found no negative perception by local communities in relation to sharing their culture and heritage sites with tourists. Perdue, Long, and Allen (1990) found that consideration for the local community's needs, and increasing community input into the development of tourism infrastructure, increases local support for the development.

Therefore, ensuring the benefits from the tourism development flow to the residents of Al-Juhfa Village is an important factor in sustainable development. The local residents require support in establishing businesses such as restaurants and hotels, and therefore, the tourism policies of the government authorities should help in establishing these and hence generate economic benefits for the community (Hampton, 2005). This will minimize economic leakages from the village as a result of tourism (Sharpley & Telfer, 2002). The Ministry of Tourism, the Saudi Commission for Tourism, and the Ministry of Culture have a role in working with tourism, travel, and Umrah companies to ensure the flow of tourist groups to the village of Al-Juhfa, hence supporting small and medium enterprises that cater them (Saufi et al., 2014).

The job opportunities for residents in tourism and commercial activities so developed will contribute to retaining tourist expenditure within the Rabigh region. This will increase the prosperity and development of the local population by increasing their income and providing services to them (Scheyvens, 2003). The potential economic loss caused to locals by tourism should also be anticipated, for example, the increase in prices for basic needs due to cost increases for goods sold to tourists. This can be prevented by establishing an equalizing mechanism for these goods and services, so that the costs for locals remain the same (Sharpley & Telfer, 2002). Thus, the negative impact of tourism activities in Al-Juhfa Village can be mitigated.

Some Al-Juhfa residents have moved to other cities to find a job because of the lack of services in the village. The findings of the study show that the provision of their basic services and needs would encourage them to stay in the village. Development of tourism in the village can provide the basis for a decent life for its residents and retain them in the local community. This will also help residents to present their culture and traditions to tourists. Therefore, developing Al-Juhfa Village for tourists can provide services that the village and neighboring areas need.

Some residents interviewed believe that tourism in the village of Al-Juhfa should not be limited to Hajjis and Umrah pilgrims but should also include non-Muslims interested in the Kingdom's civilization and history. Provision of a unique experience for international tourists would need additional training to develop the human resources of the village and equip them with the skills of dealing with such tourists and providing them with high-quality services (Scheyvens, 2003).

CONCLUSION

To optimally benefit local people, tourism development requires the adoption of sustainable principles. The residents of Al-Juhfa Village perceive tourism development as providing positive benefits. These residents can provide heritage and agricultural activities as attractions. Development of infrastructure and tourism facilities can create economic benefits and increase residents' welfare. To maximize the benefits and empower residents, there is a need to increase residents' tourism knowledge and skills so they can play an important role in its development. Such training will strengthen the support of the destination residents and sustain development, particularly within a heritage village such as Al-Juhfa.

Chapter 9

TOURISM DESTINATIONS Applying the Intertwining Model

Yoná da Silva Dalonso
University of the Region of Joinville, Brazil

Júlia Maria Lourenço
University of Minho, Portugal

Paula Cristina Almeida Remoaldo
University of Minho, Portugal

Alexandre Panosso Netto
University of São Paulo, Brazil

Abstract: This chapter presents and analyses the application of the novel version of the Intertwining Model in two tourist destinations which are strongly and successfully related to Christmas events and products in Brazil and in Finland. This analysis serves as an attempt to monitor the process of tourism development taking into account the policies implemented through time and the inter-relations between them, from the destinations' vocation for Christmas tourism. This analysis identifies stages in the evolution of public policies and their relationship to the networks of different actors, at the phases of development. This chapter confirms that as the model indicates, stakeholders have multiple roles. **Keywords**: Tourism; destination; models; public policies; stakeholders; Christmas destinations

Managing Destinations: From Theories to Practices
Bridging Tourism Theory and Practice, Volume 14, 111–126

Published under exclusive licence by Emerald Publishing Limited
ISSN: 2042-1443/doi:10.1108/S2042-144320240000014009

INTRODUCTION

Several previous studies have indicated the importance of multiple networks of stakeholders in the tourism organization process and the development of destinations (Cawley & Gillmor, 2008; Dredge, 2006; Pine & Gilmore, 1998; Saxena & Ilbery, 2008; Svensson, Nordin, & Flagestad, 2005). This chapter provides an analysis of the dynamics between the different stakeholders during the process of tourism development based on evidence from two "Christmas" destinations over two decades. Many authors have discussed the role of tourism in the development of cities, especially the process of developing public policies for those places (e.g., Arcese, Di Pietro, & Mugion, 2015; Butler, 1980, 2006a, 2006b, 2009; Elliot, 1997; Getz, 1986; Gunn, 1988, 2004; Whitford, 2011). Further, systems theory has been applied to explain how tourism works and highlights the importance of planning to improve efficiency as well as social responsibility and sustainability of a destination (Devine & Devine, 2011; Padin, 2012).

Tourism systems do not necessarily develop in a linear or predictable way, making tourism planning difficult (Farrell & Twining-Ward, 2004). The unpredictable nature of tourism as well as the failure of many models suggest the need to study a destination long-term so that a sound data base of case studies can be used for modeling (McKercher & Wong, 2004; Ritchie, 2004).

This study is based on the Intertwining Model (Dalonso, 2015; Dalonso, Lourenço, Remoaldo, & Panosso Netto, 2014), a theoretical policy model of event segment development in destinations. This model has been used in a number of studies (Blyablina, 2015; Hsu, King, Wang, & Buhalis, 2016; Montebelli & Vavassori, 2017; Rizzi & Graziano, 2017). This chapter incorporates the role of the stakeholders in the Intertwining Model based on observation of the dynamics and networking between the different stakeholders during the process of tourism development in destinations.

This model has been used to study the cities of Gramado (Brazil) and Rovaniemi (Finland) (Dalonso et al., 2014). These two cities have developed Christmas-themed tourism attractions which have become the main economic activity of these destinations and their surrounding regions (Ministry of Economic Affairs and Employment of Finland, 2017; Ministry of Tourism of Brazil, 2015). These attractions are designated "Christmas Light" (Gramado) and "Christmas in Lapland" (Rovaniemi). The Santa Claus/Saint Nicholas theme is a Christmas tradition based on various historical sources (Ger & Belk, 1996; Hall, 2008; Pretes, 1995; Tervo, Hall, & Saarinen, 2013). The cities of Gramado and Rovaniemi have developed in a similar manner (Dalonso et al., 2014). The Intertwining Model was updated by Dalonso (2015) to include dynamism in the types of functions and roles of stakeholders in each phase of the tourism destination life cycle.

This chapter provides additional findings about the importance of stakeholders, their networking relationships, and their personal interactions in the tourism development process at both destinations. These data were obtained from interviews with public institutions and private enterprise stakeholders such as local leaders, researchers, government managers, and tourism entrepreneurs in both destinations.

The results indicate that planners, public and private managers of tourism, regional organizations, and residents play distinct roles in each phase of both destinations' development. These results indicate that the Intertwining Model may be applied to destination development and emphasizes that tourism planning should connect to existing stakeholder networks and that these stakeholders may play multiple roles.

LITERATURE REVIEW

Tourism models are designed to contribute to the understanding of the various components of the tourism industry as well of the stages of planning and management and to make predictions of future trends in the sector. Tourism models can be classified as theoretical models, exploratory models, and forecast models. The development and implementation of all these models occurs at various levels: local and regional as well as national or international.

The study of the tourism destination development process has been an active research area for the last four decades (Arcese et al., 2015; Butler, 1980, 2006a, 2006b; Getz, 1986; Gunn, 2004; Hall, 2007; Lourenço, Álvares, & Danko, 2011; Ma & Hassink, 2013; Mill & Morrison, 1985, 2007; Pearce, 2008; Whitford, 2011). However few studies have included the role of stakeholders in this modeling. Recent systematic approaches that have been proposed to understand tourism's components and operation also do not include stakeholders (Akın, 2015; Garay & Canoves, 2011; Mill & Morrison, 2007).

The Tourism Area Life Cycle model (TALC) proposed by Butler (1980) has been widely cited in geography tourism studies (Gibson, 2010). This model is still used although it was originally proposed four decades ago (Butler, 2006a, 2006b). Most criticisms of this model are related to its limitations in predictability and applicability (Cooper, 1990; Garay & Canoves, 2011; Haywood, 1986; Oppermann, 1998). The descriptive destination life cycle model of tourist areas is used in a number of studies to illustrate the different phases of a destination's evolution (e.g., Agarwal, 2002; Butler, 1980, 2006a, 2006b; Cooper, 1990; Di Benedetto & Bojanic, 1993; Douglas, 1997; Getz, 1992; McElroy, 2006; Oppermann, 1998; Tooman, 1997).

Butler's model emerged during the transition from quantitative to critical geography (Coles, 2006) and became a reference model to explain the evolution of tourism destinations (Butler, 2009, 2014; Hall, 2006). The TALC model argues for the existence of identifiable stages of development, and it suggests that Butler's tourist areas follow a standard evolution represented by an "S" curve which is established by plotting the number of tourists versus time.

More recently, a number of studies have examined the role of events in the development of a destination, public policies, and local tourist planning (e.g., Getz, 2008; Roche, 1994; Stokes, 2008). As the event segment grows and becomes a key element in public policies of destinations, it is necessary to develop an effective theoretical framework to analyze the policies that are influenced by these events (Whitford, 2011). Modern society is dominated by events (Boorstin, 1992; Getz, 2007; Pretes, 1995) which are increasingly discussed using tourism experience concepts (Ek, Larsen, Hornskov, & Mansfeldt, 2008; Getz, 2012; Kim, Jun, Walker, & Drane, 2015; Lu, Chi, & Liu, 2015; Morgan, Lugosi, & Ritchie, 2010). Getz (2012) argues there will be a growing need for planners and analysts of policies with the ability to deal with the complex social, cultural, economic, and environmental issues related to the events. Events can play a specific role in attracting tourists by acting as animators and catalyzers of a positive image of the destination (Connell, Page, & Meyer, 2015; Getz, 2012).

The Intertwining Model (Dalonso et al., 2014) proposes a relationship between three main components in a destination: tourist assets (resources, equipment, and products), public policies, and stakeholders. This model (Figure 1) consists of a series of phases or temporal divisions caused by shifts

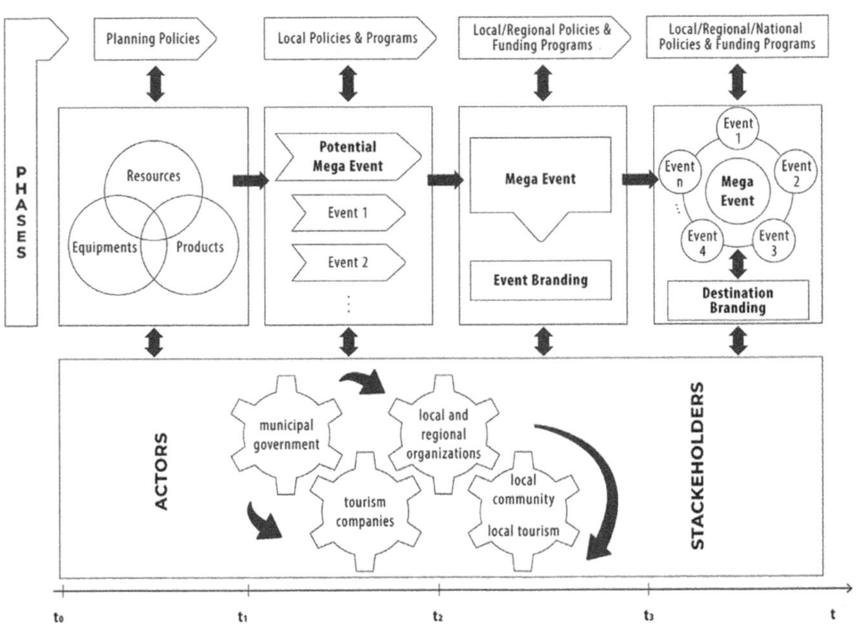

Figure 1. The Intertwining Model (Original Version).

Source: **Adapted from Dalonso et al. (2014).**

Link of original source of Figure 1: https://www.sciencedirect.com/science/article/abs/pii/S0160738314000309#f0020

in existing policies. These policies in both case studies discussed here were first enacted at the local level and afterward at higher levels: regional and national, then evolving into public funded programs.

The model illustrates this development process which starts when the destination identifies resources, equipment, and products framed within the context of local planning policies. In a second phase, these policies start to encompass other sectors besides local planning such as events planning. This phase includes the actors' perceptions or their support for the events and the arrival of tourists at the destination. A third phase begins when a mega event or attraction occurs, and for that, regional policies and funding programs have to be in place so that the event branding is successful.

The last phase involves destination branding and a myriad of smaller events or attractions developing around the mega event/attraction and for which national policies and extra funding are needed. From the analysis and application of the model, it is possible to identify the phases of the policy-making process on a timescale (t_0, t_1, t_2, t_3). The definition of the times (t_1, t_2, t_3, t_4,...) is established according to the timeline of policies and events at each destination.

The Intertwining Model examines the level of influence of events on the public policies in a destination and its interface with the different actors involved. But the stakeholders in the model are assumed to remain static across the tourism development process (Friedman & Miles, 2002). That is, the different resources, equipment, products, and policies may be different at each phase, but the actors and stakeholders were not well described. In the first version of the model, they were depicted within four wheels identified as municipal government, local and regional organizations, tourism companies, and the last one representing the local community and the local tourist. This suggests some dynamism, but this is not well defined.

In discussing and applying this model, it is necessary to consider the special characteristics of each destination and their mix of touristic services and products (Oppewal, Huybers, & Crouch, 2015). Butler (2006a) argues that each destination represents a unique combination of geography, politics economy, ethnography, and particular elements of individual behavior. Therefore, in order to apply a life cycle analysis, the destination, the tourism product, and the networks built by stakeholders should be specified.

The new version of the Intertwining Model (Dalonso, 2015) considers the four phases for all the components of the model, including the stakeholders (Figure 2). This is due to the fact that after interviewing several stakeholders, it was perceived that the roles played by different actors in the process of consolidation of the destinations branding keep evolving and enlarging as more stakeholders get involved in the tourism development process of each destination.

All this process depends on the interaction between the main stakeholders involved, often represented by municipal managers, leaders of tourism

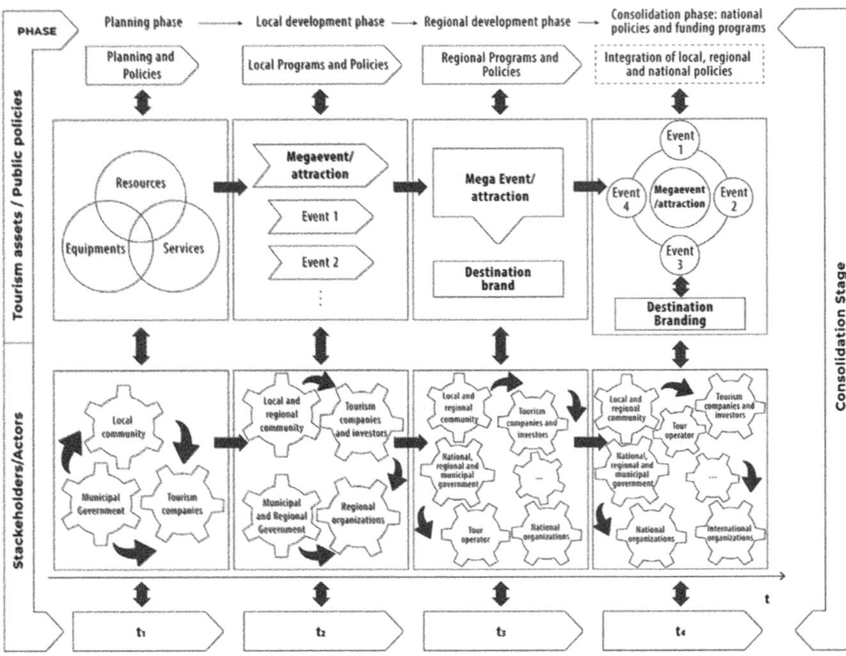

Figure 2. An Updated Intertwining Model.

***Source:* Adapted from Dalonso (2015).**

Link of original source of Figure 2: http://repositorium.uminho.pt/bitstream/1822/38439/1/TESE_Yona_da_Silva_Dalonso_18_marco_2015.pdf (page 133).

companies, local, regional, national, and international organizations, and the local community. These stakeholders play a fundamental role in the desired development of the local destination. The model allows the insertion of different stakeholders in the phases, graphically represented by circles. This new visual structure for the Intertwining Model allows the identification of the various networks established throughout the proposed phases.

The recognition of such diverse stakeholder groups and the incorporation of the perspective that their interests may not be exclusively focused in tourism activities is more realistic. Thus, this new version introduces scenarios which present higher levels of complexity in the task of understanding the roles played by different stakeholders in the multiple inter-relationships networks (Bramwell & Sharman, 1999; Cawley & Gillmor, 2008; Sautter & Leisen, 1999; Waligo, Clarke, & Hawkins, 2013).

This updated version (Figure 2) allows for future integration of concepts such as governance and leadership in the development process. It also allows the introduction of key persons who could eventually be designated as smart tourism mavericks, borrowing the concept from regional planning. These are the key actors and key stakeholders that lead the development process, inspiring and mobilizing other actors to gather behind a joint vision with prioritizations and policy mixes.

METHOD

This chapter adopts the case study method in the application of the Intertwining Model to two Christmas destinations Gramado (Brasil) and Rovaniemi (Finland). Data were collected using structured interviews with a Likert-scale used to quantify answer to questions. Interviews were conducted in 2013 and reapplied in 2014. The choice of these two Christmas destinations was due to the fact that these small destinations have successfully developed public policies toward tourism development. Secondary sources included analysis of official documents and development strategies enacted for tourism in both destinations. The Rovaniemi case covered 90 years of tourism development, while in Gramado, the period was 50 years as this second destination is a more recent one. These destinations also have one focus of tourism development allowing the research to identify the specificities of their tourism development processes in a more effective way compared to larger and more complex tourism destinations.

During the process of remodeling both Christmas tourism destinations, the primary data were obtained through nine standardized in-depth interviews of regional and local stakeholders of each destination during 2014. A questionnaire using 16 questions in nine sections was used. The sections included importance of tourism to the city (impacts), destination leadership in the regional context, active participation in tourism policies (local and regional), marketing actions and branding of the destination, interaction with the residents (public participation/actors), interaction with visitors, types of tourism business, threats to the destination, and future expectations for tourism in the destination. This division made it possible to give some coherence and homogeneity to the different interviews, allowing the comparison of the perceptions of each of the participants in the interview.

This survey was used in a pretest to five people then applied in Rovaniemi and Gramado. The goal of the pretest was to check the number and relevance of the questions used in the questionnaire, as well as to control the response time. This pretest also evaluated the accuracy, clarity, and comprehension of the questions proposed in the interview. The average interview time was 17 minutes, and all the questions presented in the survey were considered valid and relevant by respondents.

One of the criteria used to select the interviewees was that they were stakeholders directly or indirectly linked to the tourist activity of the destinations. The respondents were identified through contact with entities from both destinations, as well as the participation of researchers from the area.

For the interviews, official contacts were made by e-mail and later interviews were undertaken in loco (Rovaniemi) and by videoconference (Gramado). All interviews were recorded with the consent of the interviewees and later transcribed.

The interview questions aimed especially at Christmas events and products as well as their relevance in relation to the tourism development process of both destinations. Especially important were the answers received to questions about active participation in local and regional tourism policies such as the context for the decision-making for the tourism sector and the local community participation; the disclosing of the most important decisions across the tourism development process for the interviewees; the description of the type of interaction between the local, regional, and national government, the enterprises and the community at large; this interaction at the beginning of tourism sector in the destination cycle.

Such a method identified the perception of stakeholders regarding the historical course of development of both destinations. According to Richardson, Peres, Wanderley, Correia, and Peres (2012), a standardized interview allows greater comparability of the answers, as the respondents answer the same questions, in addition to reducing the effects and influence of the interviewer. Furthermore, the organization and analysis of data obtained in the interview is facilitated.

4. Case Studies

4.1. Gramado (Brazil)

Gramado is a city located in the south of Brazil with an estimated population of 34,110 inhabitants. The city is one of the main Brazilian tourist destinations with an economy based 90% on tourism and touristic activity. The city received 5.7 million tourists in 2012 which is an increase of approximately 17% compared with the previous year (Gramado's Tourism Board, 2013).

The "Christmas Light" event ("Natal Luz," in the original language) highlights the European characteristics of the city and is based on the Christmas traditions, brought by the German immigrants. From the first Christmas Light in the 1986, it was transformed in a 7 days event in 2014/2015. Nowadays, the festival includes 10 types of events performed for over 500 times, during a period of 73 days (Natal Luz, 2014). In 2013, the "Christmas Light" of Gramado received 1.5 million tourists compared with 5 million tourists for the whole year, having had the highest number of visitors recorded since its inception (Gramado's Tourism Board, 2013).

Gramado keeps a consolidated tourism destination in the concept of Butler's life cycle model (1980). Constraining factors such as last years' economic crisis in Brazil and the perception that Gramado is a highly priced tourism destination did not reduce tourist numbers. On the contrary, during the period 2014–2017, tourist flows increased in Gramado. Another relevant indicator, the accommodation reservations for families' 2019 summer vacation, through the most popular online rental site, shows that both cities of Gramado and Canela are most popular, surpassing small cities and coastal areas near Rio de Janeiro and São Paulo (panrotas.com, 2018).

From the analysis of the organizational dimension of tourism in Gramado, the destination presents an alignment of policies, both at the local, regional, and national levels. The organizational structure of tourism management of the municipality of Gramado is shown in Figure 3, as described by the Tourism Regionalization Program of the Ministry of Tourism of Brazil.

Gramado is the main city of the Hydrangeas micro-region and plays a strategic role in its development. The fact that the councils and entities of regional representation are based in Gramado allows the destination a competitive advantage in relation to the other cities that are part of region that is inserted. Gramado was one of the first 15 Brazilian cities to have a strategic tourism plan prepared by the Ministry of Tourism of Brazil (Ministry of Tourism of Brazil, 2010). This indicates that Gramado is a strategic market for Brazilian tourism, where the capacity of local and regional organization and articulation of the different stakeholders of Gramado are characterized as determining factors for the organization and professionalization of tourism in the region.

4.2. Rovaniemi (Finland)

Rovaniemi is located in the northern part of Finland, in the Arctic Circle, and it is the administrative capital of Finnish Lapland. It is best known as the official home of Santa Claus (Pretes, 1995) or as the home town of Santa Claus and the World Capital of Christmas (Invest in the Artic, 2009). Each year, Rovaniemi records about half a million registered visitors (Visit Rovaniemi, 2019). However, the figure does not include the significant number of unregistered overnight stays (e.g., in cottages). Around 60% of the tourists are foreigners who come to Rovaniemi in the winter season (from November till April), as this city is the main destination for Christmas tourists.

The number of foreign charter travelers has been a proxy indicator of the development of Christmas tourism in Lapland. For the first decade of the 21st century, the number of Christmas tourists in Lapland had steady around 100,000 per year, except in 2009 when in global recession period, there were about 80,000 Christmas tourists (Lappi, 2011a). The following decade, Lapland annual visitors have increased every year. This was due to the expansion of routes between Europe and Asia by the Finnish airline carrier while promoting Helsinki as a stopover hub. For example, the number of Chinese tourists who visited Lapland increased by more than fourfold

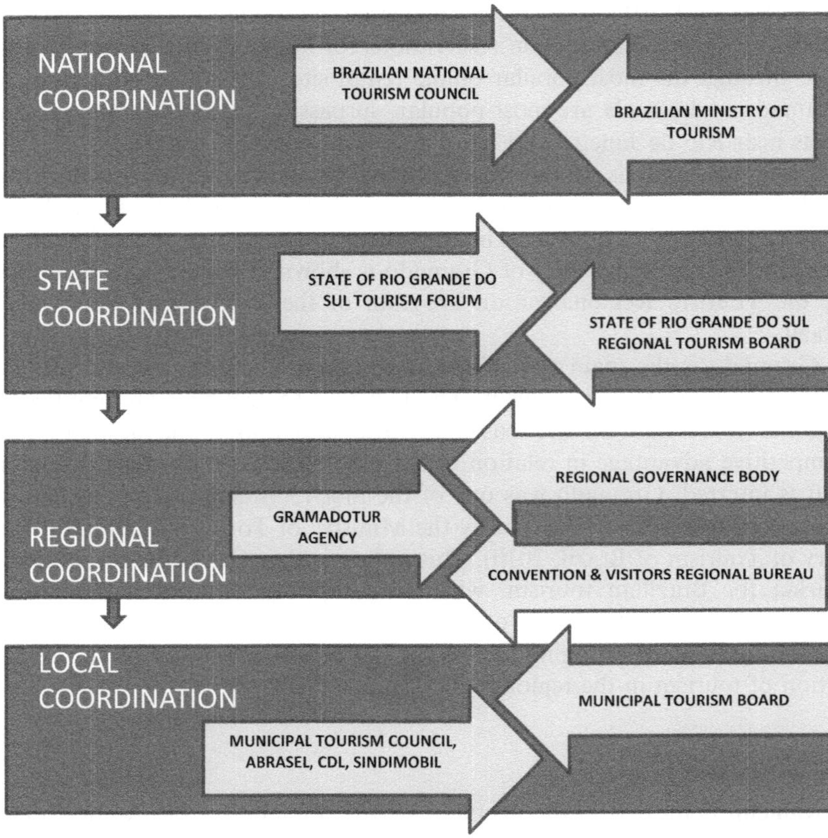

Figure 3. Actors and Stakeholders in Gramado Tourism System.
***Source:* Adapted from Hall (2007).**
Link of original source of Figure 3: https://www.taylorfrancis.com/chapters/edit/10.4324/9780203392096-22/tourism-governance-mis-location-power-michael-hall (page 248).

between 2015 and 2018, reaching 45,000. As a proxy for the attempts to reduce seasonality, statistics show that the Christmas season share of the total flights has decreased from 75% (2006) to 57% (Rintala-Gardin, 2013).

The city of Rovaniemi has built its development policies in the tourism sector around the Christmas event (Tervo et al., 2013). Its region is Finland's second largest touristic destination, after the region of Helsinki. Though Lapland represents almost a third of the country's territory, only 4% of the population resides in the region, and its economy depends on tourism (City of Rovaniemi, 2010; Haahti & Yavas, 2005).

The organizational structure of tourism in Finland (Figure 4) is led by the Ministry of Economic Affairs and Employment of Finland advised by the Finnish Tourism Board in conjunction with the Tourism Trade and other tourism organizations. In relation to regional governance, tourism policies and strategies are implemented through the articulation and involvement of different actors and stakeholders (Lappi, 2011b, 2015, 2019).

Tourism operators, tourism companies, service providers, and tourism products play a major role in structuring the policies put in place by the Finnish Tourism Board and the Lapland Regional Council. In the process of tourism development in Lapland (Lappi, 2011b, 2015, 2019), stakeholders play roles from entrepreneurs in the tourism sector to representatives of various types of organizations. Public policies are drawn up through the Lapland Regional Council (Lappi, 2011b), the regional development and territorial planning body in the Lapland Region.

The Council manages the development programs related to the national and EU structural funds. For more than a decade now, this Council has also been responsible for international cooperation efforts in the region, as well as

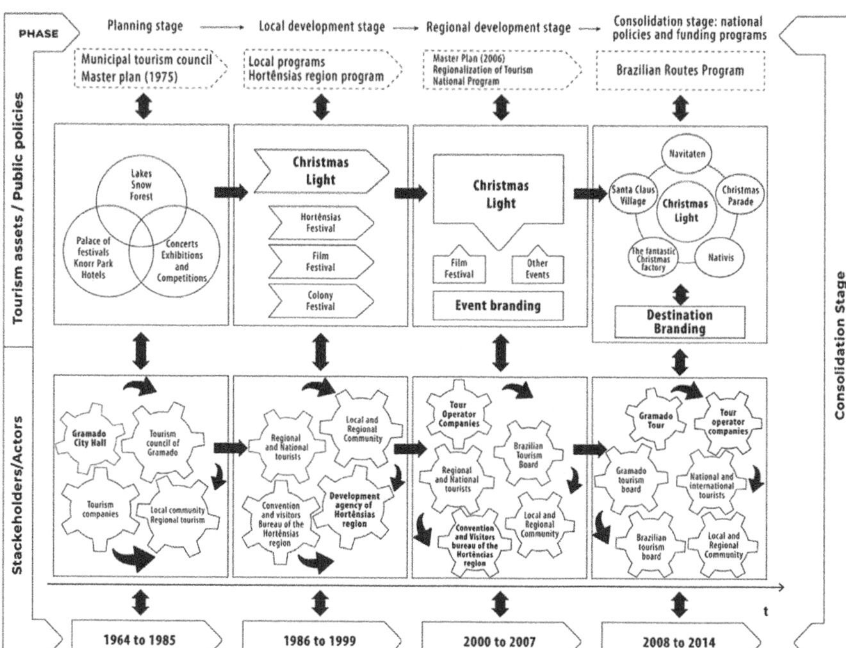

Figure 4. Application of the Intertwining Model in Gramado.

Source: **Adapted from Dalonso (2015).**

Link of original source of Figure 4: http://repositorium.uminho.pt/bitstream/1822/38439/1/TESE_Yona_da_Silva_Dalonso_18_marco_2015.pdf (page 228).

for the emergency services in the Lapland Region and for the implementation of national policy, in accordance with the Finnish regional policy legislation (Lappi, 2008). The extensive networks that have been created over decades, among key Lapland stakeholders, foster active participation in discussions on various topics of tourism competitiveness and innovation while bonding actors in collaborative efforts.

Lapland is considered an early adopter of the smart specialization concept, in this case as the Christmas tourism. The results in this region as compared to neighboring Artic regions in Sweden, Norway, Russia, and Iceland encourage regional actors to continue their efforts in applying this concept for regional innovation and development (Giacometti & Teräs, 2019).

RESULTS

The adoption of the case study method and the use of interviews in the application of the Intertwining Model to Gramado and Rovaniemi allow an in-depth identification of the tourism development phases, relating the adopted tourism policies with the tourism assets and the stakeholders' roles, during a life-cycle period of destinations. In this study, it is possible to observe the changes and the increasing number of stakeholders in the sequence of stages outlived by each destination.

Regarding the stakeholders, both destinations present very distinct characteristics over time. It follows that in Gramado (Figure 4), the process of becoming a tourism destination was initiated through a local discussion with great involvement of the residents and municipal managers (stage 1). Afterward, the destination went along a process of regionalization of the touristic product, where the regional actors led the action (stage 2). Following a trend to increasing tourists, Gramado became part of the national tourism policies promoted by the Brazilian Tourism Board, becoming a strong destination with a successful branding of events (e.g., Christmas Light event). At that stage (stage 3), Gramado became an important tourism center for the national market, attracting tour companies and operators to invest there and to commercialize its products. Through the inter-relationships built between different actors in the course of Gramado's consolidation as an important Brazilian tourism market, the destination started to brand itself internationally, through a joint action of municipal managers and government, headed by tour operators and companies (stage 4).

In the case of Rovaniemi (Figure 5), the residents and private initiatives (especially tour operators) played an important role promoting the Christmas branding. However, the strategic actions of the Rovaniemi Tourism Board

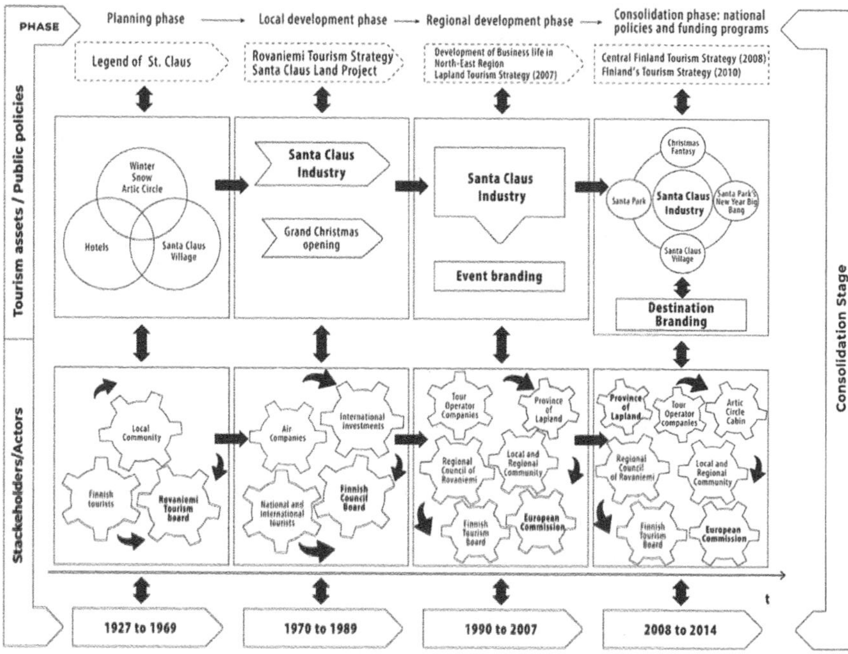

Figure 5. Application of the Intertwining Model in Rovaniemi.
Source: **Adapted from Dalonso (2015).**
Link of original source of Figure 5: http://repositorium.uminho.pt/bitstream/1822/38439/1/TESE_Yona_da_Silva_Dalonso_18_marco_2015.pdf (page 240).

were crucial in attracting national tourists to the destination (stage 1). In the next stage (stage 2), Rovaniemi became the focus of international capital, through investments made by airlines and international companies. This resulted in a growth of international tourism flows that were stimulated by the Finnish Council Board. The grand opening promoted by the Christmas industry for the international investors led to the intensive networking between companies and municipal, regional and national managers. This dynamism culminated in capturing investments from the European Union (stage 3). Therefore, Rovaniemi became an important center for tourism at the Lapland Region (stage 4), consolidating itself as a strategic tourism market in the European Union.

In the case of Gramado, the tourism development process started much later when compared to Rovaniemi, but both destinations followed the same stage development path until consolidation. The Christmas Light event was based on different events developed in the destination over four decades of product development. However, Christmas tourism consolidated Gramado as an important Brazilian destination. Despite the crisis faced by the Brazilian tourism market in recent years, Gramado has managed growth above the national average, remaining in the phase of tourism consolidation.

Rovaniemi, after nine decades of tourism development, has already established itself internationally as the key destination brand in the Santa Claus industry, receiving the title of "Official Hometown of Santa Claus ®." Rovaniemi, likewise, has been showing a recent growth in tourist demand in recent years, especially motivated by winter tourism and the snow phenomenon (Falk & Vieru, 2019; Lapland Chamber of Commerce, 2019; Varnajot, 2019).

The tourism industry stakeholders played a more important role in Rovaniemi than in Gramado, but public policies and funding, respectively from the European Union and from Brazil, were key to foster this development. Nevertheless, it is important to consider that both destinations present peculiarities in the formatting of its tourism product (e.g., Christmas celebration). The application of the model to other destinations will depend on the existence of a tourism brand, translated in the public policies and its networking inter-relationships, at different stages of their development process, with several stakeholders, identified as strategic elements in the consolidation stage of the model.

The interviews reveal the importance of tourism for the development of both destinations, as well as similarities in their tourist development path. The questions about interaction with visitors and active participation of locals produced the same results in Rovaniemi but not in Gramado. This leads us to doubt if the tourism strategies are so widely shared in Gramado as in Rovaniemi. This later destination keeps producing 3 years Strategic Frameworks, while Gramado does not. It may be that Gramado relies more on tourism mavericks than Rovaniemi. More research is needed here. One of the similarities is that both destinations have developed around the Christmas theme. Another important factor observed in the interviews was the integration and regional promotion capacity of the analyzed destinations. Both Gramado and Rovaniemi presented themselves as the driving force behind tourism in the region in which they are located. In both destinations, there was a strong economic dependence on tourism.

Regarding the stakeholders, the application of the Intertwining Model demonstrates a strong inter-relationship between the planners, public and private tourism managers, regional organizations, and the local population, all playing distinct roles in each phase of both destinations' successful tourism development process.

CONCLUSIONS

The use of models, such as the Intertwining Model, serves as an attempt to monitor tourism development processes through the policies implemented across time and the inter-relations that happen on the way. The application of the model indicates that Gramado and Rovaniemi displayed similar behaviors in their development process and evolution. With that, the application of the proposed model brought to light its practical function, approaching the theory to the practice of tourism. It is relevant to consider that both Gramado and Rovaniemi own characteristics to the consolidation of its touristic brand (Christmas tourism).

Regarding stakeholders, the model has shown a strong inter-relationship between planners, public and private managers of tourism, regional organizations and residents, playing distinct roles in each phase of the destinations applied. The proposed model works on the assumption of the need of public participation under the scope of local, regional, and national policies, which elevate its application to the destination as well as other models that deal with an analysis of the community involvement under the regional scope (e.g., Whitford, 2011). The analyses of networks allow better identification of those involved and a more heterogeneous reading of stakeholders in the destination (Cooper, Scott, & Baggio, 2009).

Thus, the application of the Intertwining Model reveals it is effective in destinations with characteristics like Gramado and Rovaniemi. Its application to other destinations will depend on the existence of a strong tourist vocation, translated in the public policies at the destination. The model does not include an assessment of these periods in destination tourism, as the phase of stagnation and decline phase, proposed in other studies (e.g., Butler, 1980). In this sense, its application was limited to destinations that want to develop themselves as successful destinations, through the consolidation of established public policy, derived from established tourist assets and the inter-relations between different stakeholders by long periods.

Therefore, a constant evaluation of the destinations is needed to identify the period of permanence in the consolidation phase. Problems lie ahead such as climate change with more no-snow months and an increased carbon footprint due to more transportation such as flights for Rovaniemi and car travels for Gramado, the changes from booking through traditional tour operators to online individual booking, the expected mass tourism littering of the pristine wilderness and neat places, the potential noise and mess caused by massive online booking. However, interviews with the inhabitants of both locations reveal that they have a long-time relationship with tourists, and they consider them part of their cities as the income generated by tourism is much needed. Therefore, collaborative not confrontational strategies are envisaged to solve the problems ahead.

This evolution of Intertwining Model allowed a more analytical reading of the role that stakeholders take at different stages of development of tourist destinations studied. On the first version (Dalonso et al., 2014), the stakeholders remained static along the development process, while in this new proposed version, the model identifies the networks that were established from the input and output of different actors. This occurs at different times and not does follow necessarily a logical order. The application of the model to the destinations proved these dynamics. Another aspect that stands out from this new version perspective is to identify stages in the evolution process of public policies, represented by different phases: planning, local development, regional development, and consolidation phase.

Finally, the new version graphically highlights the application of the model to the consolidation phase. As the Intertwining Model is applied to successful destinations, any factor that does not allow its applicability in the fourth period proposed may place at risk the consolidated destination condition. This can be read as a visual warning about the way that tourism planning connects to existing networks. Finally, it is important to consider that exploratory models are instruments that propose a much more assertive than inductive analysis of a specific tourism destination life cycle.

Chapter 10

RESIDENTS AND TOURISM COVID-19 Evidence, Cape Verde

Edson Santos
University of Algarve, Portugal

Patrícia Pinto
University of Algarve, Portugal

Luís Nobre Pereira
University of Algarve, Portugal

Abstract: This chapter investigates residents' perception of tourism development on the small tourist island of Santiago, Cape Verde, during the crisis caused by the COVID-19 pandemic. Specifically, it shows how the pandemic situation is affecting residents' perceptions of tourism impacts, their quality of life, and support for tourism development. Data were collected during winter 2020–2021. Partial least square structural equation model (PLS-SEM) was used to assess interrelations among constructs. The study contributes to the literature by extending the understanding of residents within the current COVID-19 pandemic context. Results demonstrate that their support for tourism was positively influenced by their quality of life. **Keywords**: Residents' perception; support for tourism; quality of life; COVID-19; Cape Verde; tourism impact

INTRODUCTION

In 2019, there were 1.4 billion international arrivals globally (+5% from 2018), generating US$1.7 trillion (+4% from 2018) in exports (UNWTO, 2020b). However, exponential growth in global tourism came to a halt due to

Managing Destinations: From Theories to Practices
Bridging Tourism Theory and Practice, Volume 14, 127–139

ISSN: 2042-1443/doi:10.1108/S2042-144320240000014010

the novel SARS-CoV-2, a respiratory disease responsible for COVID-19 pandemic crisis (Uğur & Akbıyık, 2020). Previous studies that investigated global pandemics and their ripple effects on international tourism industry provide insight into the economic, social, and health repercussions caused by deadly disease outbreaks (Tisdell, 2020). The effects of the current pandemic have devastated developing economies (Gössling, Scott, & Hall, 2020), especially vulnerable small islands highly dependent on external revenue from tourism-related activities such as Cape Verde (Ribeiro, Woosman, & Megeirhi, 2020). In 2019, tourism contributed 37.2% (direct, indirect, induced) to Cape Verde's gross domestic product (GDP) and 39.3% to employment (WTTC, 2020a, 2020b). However, due to the closure of the country's leading tourism markets (European countries), Cape Verde's economy has been seriously impaired by the pandemic.

The economic, social, cultural, and health consequences from COVID-19 have affected residents' well-being and quality of life (QoL) in several destinations (Ramkissoon, 2020b). The local community's QoL is important because it has the potential to influence residents' attitude and consequently their support for tourism development (Nunkoo & Ramkissoon, 2010b). In addition, there is evidence that residents satisfied with their QoL are more likely to support tourism development (Uysal, Sirgy, Woo, & Kim, 2016). Therefore, it is important to understand residents' well-being and QoL in a crisis situation (Ramkissoon, 2020b).

Although the studies mentioned above have contributed significantly to the body of knowledge of residents' perception, QoL and support for tourism, studies examining these constructs within the current pandemic situation are limited. To address this gap in the literature, this chapter aims to investigate how the current COVID-19 pandemic has affected residents' perception of economic, social–cultural, and environmental impacts of tourism, and how this influences their QoL and consequently their support for tourism development. Using social exchange theory (SET) as the theoretical base, this study measured these constructs within a small island developing destination context.

RESIDENTS AND TOURISM

Resident support for tourism is an important factor to consider for the development of sustainable tourism (Gursoy, Ouyang, Nunkoo, & Wei, 2019; Yu, Cole, & Chancellor, 2018). Understanding the antecedents of residents' support for tourism development is vital for the success of a destination (Gursoy & Rutherford, 2004). Several studies presented in the literature have been developed under the SET framework to explain residents' perception of tourism impact as the primary predictor that influences them to support tourism development (Nunkoo & Gursoy, 2012; Nunkoo &

So, 2016; Stylidis, Biran, Sit, & Szivas, 2014). Based on the SET framework, Nunkoo and Ramkissoon (2011) proposed a model that investigated how residents' perceived benefits, costs, and overall community satisfaction are influenced by trust in tourism institutions, power to influence tourism, neighborhood conditions and consequently contribute to their support for development. Nunkoo and Gursoy (2012) examined identity through the lens of resource-based occupation, environment and gender, in an island community, and how these constructs influenced residents' support for tourism development. Later, social–cultural, environmental, and economic perceptions were considered to explain residents' support for tourism development in Stylidis et al. (2014).

Within a tourism community context, residents' QoL has also been considered a significant predictive factor of support for tourism development (Campon-Cerro, Folgado-Fernandez, & Manuel, 2017; Eslami, Khalifah, Mardani, & Dalia, 2019; Nunkoo & So, 2016). Nunkoo and So (2016) developed a baseline model of residents' support and compared it with four competing models. Although their study found residents' perception of positive impact to predict support for tourism, they did not prove a relationship between residents' satisfaction with the QoL and support for tourism. Other studies demonstrated that residents' perception, community attachment, and overall QoL are significant predictors for their support for tourism development (Campon-Cerro et al., 2017). Eslami et al. (2019) measured community attachment within the triple bottom line (economic, social, and environmental) impact to examine how residents perceived their QoL and its influence on their support. Despite these valuable contributions, studies that examine these constructs within a crisis situation are limited. Understanding how residents perceive their QoL in the context of the current COVID-19 pandemic is vital for destination managers who seek to implement strategies that influence residents' support for future tourism development. Based on this coverage, the following hypothesis is proposed.

H1. Residents' QoL influences their support for tourism development.

Tourism Impact and Quality of Life

Bearing in mind that the influx of tourism in a community may significantly affect all aspects of residents' lives (Woo, Uysal, & Sirgy, 2018), it is important to take into account how they perceive the economic, social–cultural, and environmental impacts of tourism development (Nunkoo & Ramkissoon, 2011). Early studies reported that residents favored economic and cultural benefits, while reluctant of the social and environmental impact of tourism development (Lui & Var, 1986). Using Butler's (1980) hypothesis theory, Akis, Peristianis, and Waner (1996) examined the social, economic, and environmental status of the host community and the perceived impact of tourism development. They confirmed that residents who were more exposed to tourism demonstrated pro-tourism behavior.

As a means to strengthen the residents' perception and support for tourism debate and to answer calls for theoretical support, scholars have adapted the SET to explain better these reactions within the tourism community (Ap, 1992; Gursoy, Ouyang, Nunkoo, & Wei, 2019; Hadinejad, Moyle, Scott, Kralj, & Nunkoo, 2019; Nunkoo, Smith, & Ramkissoon, 2013; Rasoolimanesh & Seyfi, 2020). The SET provides an understanding of residents' perception and attitude by explaining the relationship between the economic, social, and environmental benefits and costs associated with tourism development (Perdue, Long, & Allen 1987). Using SET, Jurowski, Uysal, and Williams (1997) proposed a model to better understand residents' perception of these impacts by examining how economic gains, resource use, community attachment, and attitude toward preservation of natural environment influence residents' reactions to tourism. The underlining principles of SET are grounded on human behavior and social interaction within an exchange setting (Nunkoo, 2016). Favorable perception is revealed if individuals trust that benefits exceed costs in the exchange process (Látková & Vogt, 2012). Moreover, residents who perceive tourism development positively and are conscious of its contributions to their QoL, as a consequence of the social exchange, support further development (Andriotis & Vaughan, 2003).

Residents' perception of economic, social–cultural, and environmental impacts from tourism can influence their QoL (Nunkoo & Ramkissoon, 2010a). Andereck, Valentine, Vogt, and Knopf (2007) examined perceived negative impact, positive economic, social, and environmental impacts and their effect on residents' QoL. Other studies attempted to understand residents' economic, social, cultural, and environmental perceptions of tourism impact and its influence on residents' well-being and QoL (Kim, Uysal, & Sirgy, 2013). A recent study measured the economic, social, and environment constructs as predictors of residents' QoL and support for sustainable tourism development (Eslami et al., 2019).

The centrality of studies has contributed to the understanding of residents' perception of tourism impacts and its effect on their QoL. However, few studies have examined this subject in the context of a pandemic, hence these hypotheses as illustrated in Figure 1.

H2. Residents perceived economic impact of tourism influences their QoL.

H3. Residents perceived sociocultural impact of tourism influences their QoL.

H4. Residents perceived environmental impact of tourism influences their QoL.

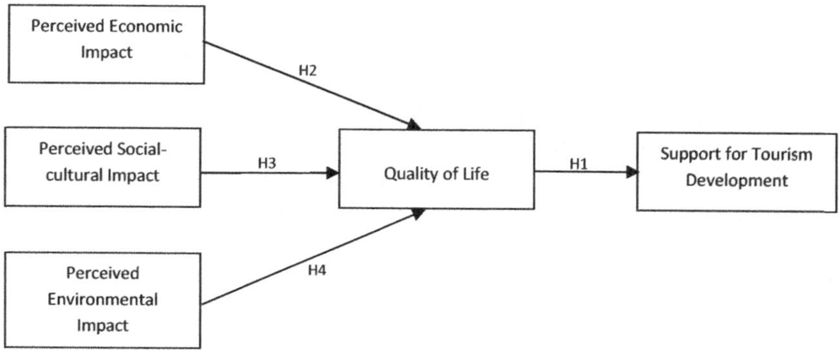

Figure 1. Conceptual Model and Research Hypotheses.

Study Methods

The research setting of this study is the Island of Santiago, Cape Verde. Cape Verde is a constellation of 10, nine with inhabitants, small islands (Boa Vista, Brava, Fogo, Maio, Sal, Santa Luzia, Santiago, Santo Antao, Sao Nicolau, Sao Vicente) located about 500 km of the West Coast of Africa, neighboring Senegal, Gambia, and Mauritania (Eusébio, Lopes, & Carneiro, 2016; Santos, 2014; Santos & Rodriguez-Campo, 2014). The majority of the population, about 88%, is concentrated on four islands of Santiago (55%), Sao Vicente (15%), Santo Antao (9%), and Fogo (8%) (WorldBank, 2018). About 28% of the population of Santiago resides in the capital, Praia (WorldBank, 2018).

The country's geopolitical position plays a vital role in the socioeconomic and cultural development (Santos, 2014), thus, contributing significantly to its tourism development, which commenced in the early 1960s (Santos & Rodriguez-Campo, 2014). The absence of political unrest, natural disaster, and religious and ethnic conflict make Cape Verde an attractive tourism destination (Lopes-Guzman, Borges, Hernandez-Merino, & Cerezo, 2013). Over the years, tourism has been positioned at the top of the political agenda as an engine for economic growth (Ribeiro, Pinto, Silva, & Woosnam, 2017). Although all of Cape Verde's nine inhabited islands possess unique characteristics and are considered tourism destinations, most of the tourism activity concentrates on the islands of Sal, Boa Vista, Sao Vicente, and Santiago (Mitchell & Li, 2016). Tourism accommodations on these islands represent 90.7% of the total (Sal – 45.4%, Boa Vista – 30.4%, Santiago – 8.8%, and Sao Vicente – 6.1%) (INE, 2020). Official data also report that international tourism arrivals reached 819,308 (INE, 2020), with a total tourism receipt of \$502 million in 2019 (UNWTO, 2020b). World Travel and Tourism Council data reveal that tourism contributed 37.2% to the GDP and employed 39.3% of the population (1 of 3 jobs) in 2019 (WTTC, 2020).

As an attempt to better understand residents' support for tourism during the COVID-19 pandemic on the island of Santiago, a survey instrument was designed to collect data. The instrument was divided into three parts to gather data regarding demographic background, perception of QoL, and support of residents for tourism development. The demographic section of the questionnaire was intended to gain insight on their age, gender, education, profession, and so on. A 5-point Likert scale (1 = strongly disagree and 5 = strongly agree) aimed to understand residents' perceptions of tourism impacts (economic, sociocultural, and environmental), while another 5-point Likert scale (1 = not affected and 5 = very affected) gathered data regarding their perception of QoL and support for tourism development. The items used to measure the constructs are listed below and were chosen considering the studies of Eslami et al. (2019) and Kim et al. (2013). The questionnaire was administrated in English and Portuguese languages. Before applying the survey, a pretest that included 20 residents of Santiago was conducted to understand question comprehension and clarity. Results from the pretest confirmed that the level of comprehension was satisfactory.

To ensure familiarity with the change regarding economic, sociocultural, and environmental perception about tourism impact, the population included residents aged 18 or older. The majority of data were collected in the capital Praia. Questionnaires were sent via social media and email platforms during December and January 2020–2021. Out of the 230 questionnaires distributed, 114 were returned and considered valid. In this study, the items used to measure residents' perception of tourism impacts and their QoL, as well as their support for tourism during a pandemic, did not follow a normal distribution (univariate tests of normality: $p = 0.000$) (Hair, Risher, Sarstedt, & Ringle, 2019).

A descriptive analysis using the software IBM SPSS Statistics 26 was used to examine participants' profile. In addition, a partial least square structural equation modeling (PLS-SEM) was applied to evaluate the conceptual model using the statistical analysis software SmartPLS 3 (Ringle, Wende, & Becker, 2015) to test the proposed model and hypotheses represented in Figure 1. Studies suggest that absence of the normality distribution assumption as an important reason for using PLS-SEM for construct measurement compared to the covariance based-SEM approach (Do Valle & Assaker, 2016). Applying PLS-SEM, the measurement model was first assessed, followed by the evaluation of the structural model.

Study Findings

The sample shown in Table 1 includes female (58.8%) and male (41.2%) residents, age 18 or more, but the majority of residents were 31–45 old (47.4%), followed by 18–30 (34.2%). Single residents represented 68.4%, and were educated with undergraduate (48.2%), masters (28.1%), and PhD

Table 1. Demographic Profile of the Sample.

Variables (n = 114)	n	Percentage
Gender		
Male	41	41.2
Female	67	58.8
Age		
18–30	39	34.2
31–45	54	47.4
46–50	8	7.0
>50	13	11.4
Marital Status		
Single	78	68.4
Married	36	31.6
Education		
High School	5	4.4
Undergraduate	55	48.2
Masters	32	28.1
PhD	22	19.3
Occupation		
Student	28	24.6
Public sector	75	65.8
Private sector	7	6.1
Self-Employed	4	3.5
Income		
CVE20.000–40.000	35	30.7
CVE40.000–60.000	5	4.4
CVE60.000–80.000	7	6.1
>CVE80.000	67	58.8
Length of Residence		
4–5 years	21	18.4
>5 years	93	81.6

(19.3%) degrees, respectively. Public sector workers corresponded to 65.8%, while only 6.1% worked for private enterprise and 3.5% were self-employed. The remainder 24.6% residents were students. Majority (58.8%) had a total household income of €725.52 (CVE80.000) or more, while 30.7% received a salary between €181.38–362.76 (CVE20.000–CVE40.000). More than 4/5th of residents (81.6%) lived more than 5 years in Santiago, 18.4% for 4–5 years.

Measurement Model Assessment. The measurement model of the study examined the reliability and validity of each construct. As observed in Table 2, with the exception of two items (Overall community life: loading = 0.698 and "Tourism contributes to social inclusion": loading = 0.649), all of the items included in the model have loadings above the recommended threshold of 0.70 (Chin, 2010). These results confirm that the individual reliability of each item to measure the latent constructs are well within the standardized loading factor since Sarstedt, Ringle, Smith, Reams, and Hair (2014) note that loadings that exceed 0.70 indicate explain over 50% of the indicator's variance. With respect to the construct reliability, the composite reliability coefficients were considered. As shown in Table 2, the composite reliability values ranged from 0.809 (Support for Tourism Development) to 0.933 (Perceived QoL), both exceeding the recommended threshold of 0.70 (Hair et al., 2019). As observed by Ringle, Sarstedt, Mitchell, and Gudergan (2020), the internal consistency reliability results should ensure values higher than 0.70 and below 0.95.

The next stage of model assessment consisted of the convergent and discriminant validity examination. Convergent validity was used to "measure the extent to which a construct converges in its indicators by explaining the items' variance" (Sarstedt et al., 2014, p. 108). This verification was done using the average variance extracted (AVE). Its values shown in Table 2, ranging from 0.516 to 0.823, indicate that the construct explains at least 50% of the item variance (Ringle et al., 2020). The discriminant validity, which followed Fornell and Larcker's (1981) criterion, sought to compare each construct's AVE value with the squared inter-construct correlation (a measure of shared variance) of that construct with all other constructs in the structural model (Hair et al., 2019; Sarstedt et al., 2014). Table 2 shows that the square root of AVE of each construct exceeds the recommended threshold. Discriminant analysis evaluation followed the heterotrait-monotrait ratio (HTMT) examination to verify if the measured constructs are empirically distinct. As observed in Table 3, none of the HTMT values exceeds the 0.90 threshold, and therefore, they are considered acceptable (Voorhees, Brady, Calantone, & Ramirez, 2016).

Structural Mode Assessment. Upon satisfactory confirmation of the measurement model, the study proceeded to evaluate PLS-SEM results with the assessment of the structural model (Hair et al., 2019). This technique was used in this study as an attempt to find evidence that supports the theoretical model to assess the relationships between constructs (Avkiran, 2018) and empirically test the hypotheses model (Hair, Ringle, & Sarstedt, 2011). As observed in Table 4, the study evaluated the relationship between residents' economic, sociocultural, and environmental perception and QoL, and how

Table 2. Results of Measurement Model.

	Means	Loading	CR	AVE	*t*	Significance
Perceived Economic Impact			0.910	0.669		
My monthly income	0.742	0.759			9.080	0.000
My current job	0.770	0.789			9.312	0.000
My employment opportunity	0.800	0.811			12.108	0.000
My cost of living	0.861	0.857			21.215	0.000
My monthly budge	0.868	0.867			24.270	0.000
Perceived Sociocultural Impact			0.847	0.581		
The people in the community	0.158	0.751			4.747	0.000
Overall community life	0.165	0.698			4.243	0.000
My leisure activities	0.081	0.830			10.222	0.000
Cultural activities	0.107	0.765			7.154	0.000
Perceived Environmental Impact			0.822	0.698		
Conditions of the community environment	0.079	0.822			10.405	0.000
Environment cleanliness	0.057	0.849			14.982	0.000
Perceived Quality of Life			0.933	0.823		
My life as a whole	0.012	0.945			78.725	0.000
The way I spend my life	0.021	0.928			43.625	0.000
The way I feel about life	0.040	0.845			21.013	0.000
Support for Tourism Development			0.809	0.516		
Tourism development is very important for Cape Verde	0.111	0.745			6.732	0.000
Tourism contributes to the disseminations of our cultural heritage	0.160	0.712			4.465	0.000

Table 2. *(Continued)*

	Means	Loading	CR	AVE	*t*	Significance
Tourism provides economic opportunities	0.123	0.761			6.190	0.000
Tourism contributes to social inclusion	0.135	0.649			4.789	0.000

Note: CR = composite reliability; AVE = average variance extracted.

Table 3. Correlation Among Latent Variables.

	(1)	(2)	(3)	(4)	(5)
(1) Perceived Sociocultural Impact	0.76[a]				
(2) Perceived Economic Impact	0.14|0.19[b]	0.82[a]			
(3) Perceived Environmental Impact	0.52|0.80	0.21|0.28[b]	0.84[a]		
(4) Perceived Quality of life	0.38|0.40	0.37|0.39	0.42|0.58[b]	0.91[a]	
(5) Support for Tourism Development	0.24|0.38	0.10|0.13	0.13|0.22	0.27|0.32[b]	0.72[a]

[a]Diagonal values correspond to the squared root value of average variance extracted (AVE) for each latent variable to assess the Fornell-Larcker criterion.
[b]Heterotrait-monotrait ratio (HTMT) values.

Table 4. Hypothesis Testing Results.

Hypotheses		Path Coefficients	*t* Statistic	Significance	Results
H1	QoL → STD	0.268	3.194	0.001	Supported
H2	PEcI → QoL	0.294	4.036	0.000	Supported
H3	PSCI → QoL	0.216	2.392	0.017	Supported
H4	PEvI → QoL	0.242	2.238	0.025	Supported

Note: R^2 for QoL = 0.293; R^2 for STD = 0.072. Q^2 of Cross-Validity Redundancy: Q^2 for QoL = 0.225; Q^2 for STD = 0.028.

this influenced residents' support for tourism during COVID-19 pandemic. Results in Table 4 show that the four hypotheses are supported because the corresponding path coefficients are statistically significant (all $p < 0.05$).

Moreover, the path estimates are all positive, meaning that the relations between the constructs are positive. The coefficients range from 0.216 (for *H3*, i.e., the relation between Residents' Perceived Economic Impact and QoL) and 0.294 (for *H2*, i.e., the relation between Residents' Perceived Sociocultural Impact and QoL).

The next step evaluated the R^2 of each endogenous construct that measures the variance explained by the model and is thus a measure of the model's predictive accuracy (Sarstedt et al., 2014). As presented in the notes of Table 4, the values obtained for R^2 were QoL = 0.293 and STD = 0.071, meaning that the model is better to explain QoL than STD. This assessment was followed by the evaluation of the Q^2 obtained by the blindfolding procedure. Scholars suggest that the smaller the difference between predicted and original values, the greater the Q^2 and thus the model's predictive accuracy (Sarstedt et al., 2014). For adequate predictive accuracy, the Q^2 values should be positive. The Q^2 values obtained in this study are as follows: QoL = 0.225 and STD = 0.028.

CONCLUSION

This study examined how residents perceived the economic, sociocultural, and environmental impact of tourism and its influence on their QoL and support for tourism development on the Island of Santiago, Cape Verde. Building on previous literature, the current study attempted to contribute to the body of knowledge by extending the understanding of residents' attitude to support tourism within a crisis situation in a small island context. Studies have proposed that research is needed to better understand the influence of residents' perception of QoL on their support for tourism development in different settings (Hanafiah, Jamaluddin, & Zulkifly, 2013). Results from *H1* show that residents' support for tourism was positively influenced by QoL. This finding is in line with previous studies that found residents' QoL to be a significant predictor of support for tourism development (Campon-Cerro et al., 2017). Similar results were found by Woo, Kim, and Uysal (2015) who considered residents' overall QoL an effective predictor for support for further tourism development.

Using SET as theoretical support, this chapter evaluated residents' economic, sociocultural, and environmental perception of tourism impact and influence of QoL and support for tourism. SET suggests that residents' perceive tourism development based on the exchange of benefits and cost associated with the economic, social, and environmental impact of tourism (Nunkoo & So, 2016). "By supporting the industry, residents try to equalize or ensure a balance in their exchange" (Nunkoo & Ramkissoon, 2012, p. 1011). Results of the study reveal that residents' perception about tourism impact positively influenced their QoL (*H2–4*). These results are consistent with Campon-Cerro et al. (2017) findings that reported positive relation of residents' economic, social, and environmental benefits to enhance residents' QoL. Others also found that positive perception of sociocultural impact positively affected residents QoL (Yu et al., 2018). More recently, Eslami et al. (2019) confirmed a positive and significant relationship between residents' economic, social perception, and tourism environmental impacts and QoL.

Similar to other studies, there are theoretical and practical implications worth mentioning. The research corroborates previous investigation demonstrating that residents' support for tourism development is influenced by their QoL. Economic, sociocultural, and environmental constructs have been examined in several settings to better understand their influence on residents' QoL and support for tourism. This chapter extends the literature to a small island context in a crisis situation, i.e., the current COVID-19 pandemic.

The research contributes to the understanding within a small island tourism destination of the effects that the current COVID-19 pandemic had on residents' economic, sociocultural, and environmental perception and how this influenced their QoL and support for tourism development. Likewise, the study also attempted to provide insight on how these measures should be considered to reposition future tourism planning and development toward a more sustainable direction. Ramkissoon (2020a) notes that a thoughtful approach of residents' concerns that governs proper management strategies to promote sustainable behaviors is fundamental for the future of tourism. As literature suggests, residents' perception and attitude toward tourism are important for destinations, especially those dependent on tourism. Therefore, it is important to address concerns pertaining to the local community and involve them in future planning.

With regards to the practical implication, while in 2019, Cape Verde was on course to accomplish its long-term goal of one million annual international tourists' arrivals, the pandemic brought the industry to a halt. Eventually, travel restrictions will be lifted and the pandemic will fade away (Kock, Norfelt, Josiassen, Assaf, & Tsionas, 2020). Therefore, the responsibility lies on the decision-making of regional and local authorities to implement long-term strategies toward a more sustainable approach in the communities. Community engagement and collaboration between the private–public sector, community members, and government are perhaps a path toward accomplishing the Sustainable Development Goals. The current crisis has provided the opportunity to reflect on collective efforts that benefit all stakeholders and guide tourism communities in the direction of sustainability.

As the level of uncertainty increases, the long road to recovery is heavily hinged on effective decision-making and long-term sustainable strategies to maintain residents' support for tourism development in the community (Yeh, 2020). Therefore, an in-depth understanding of proper management approach that involves residents in tourism-related activities to adapt supportive behavior is important for the future of tourism in Cape Verde. Furthermore, crises such as the novel COVID-19 pandemic provide an opportunity for the government of Cape Verde to create leadership programs across varying groups (family, community, private–public), aiming to coordinate and implement collective efforts toward a more sustainable approach for development. Considering the importance of tourism to the socioeconomic development in this island nation, these efforts, which aim to benefit communities, will gain extra support from residents for future development. Additionally, dissemination of information guided by coordinated efforts between Cape Verde's authorities and international tour operators is vital for the restart of tourism. Complementary efforts from managers and government should focus on the creation of mechanism to empower residents with proper guidelines to interact with tourists (safety behavior norms, which can gain tourists' trust and make them feel secure during their visit to the islands) (Kamata, 2021). Equally important, established guidelines for responsible tourist behavior, inclusive management system, and benefit-sharing could also be adapted as standard approaches for tourism recovery (UNWTO, 2020a).

As with any study, there are important limitations worth mentioning. First, data for the study were collected in only one city (the Capital of Cape Verde, Praia). In order to gain a better understanding of residents' perception and attitude, it is suggested that data be collected on other regions and islands, especially the two most touristic islands of Cape Verde (Sal and Boa Vista). A second limitation is related to the sample size and profile. The sample is small, and the majority of the participants consisted of highly educated residents and students. Inputs from residents with diverse background and different demographics are recommended for future research.

Furthermore, the study proposed a model to examine residents' economic, sociocultural, and environmental perception as well as QoL as an antecedent for support for tourism development. Further research should include other constructs in the model, such as community attachment, residents' empowerment, and involvement in tourism activities. Finally, data for this investigation were collected during the present COVID-19 pandemic. Studies that report resident perception and attitude before, during, and after the pandemic would be interesting. In addition, a longitudinal research that examines resident perception for a period of time would also contribute to a better understanding of the factors affecting residents' support for tourism development.

Acknowledgments—The study was financed by National Funds provided by FCT-Foundation for Science and Technology through project UIDB/04020/2020.

Chapter 11

HERITAGE CITY IMAGE
Local Players' Perceptions

Miriam Mota
University of Algarve, Portugal

Bernardete Sequeira
University of Algarve, Portugal

Manuela Guerreiro
University of Algarve, Portugal

Patrícia Pinto
University of Algarve, Portugal

Abstract: Although tourism destination image is a widely studied subject, the perspective of local players is generally neglected, albeit its relevance for informing the positioning and brand management strategies of the places is recognized. This chapter aims to determine the perceptions of key local public organizations from the historical-cultural and heritage sectors and companies linked to commerce and tourism (private sector) about the historic center of a United Nations Educational, Scientific, and Cultural Organization (UNESCO) World Heritage site in Brazil. The results of this investigation contribute to the development of marketing and tourism development strategies in historic towns, especially those classified as World Heritage by UNESCO. **Keywords**: UNESCO; World Heritage Site; destination image; destination; city center; city image

Managing Destinations: From Theories to Practices
Bridging Tourism Theory and Practice, Volume 14, 141–157

ISSN: 2042-1443/doi:10.1108/S2042-144320240000014011

INTRODUCTION

In order to protect and preserve cultural and natural heritage, the United Nations Educational, Scientific and Cultural Organization (UNESCO) created the World Heritage Convention in Paris in 1972. The World Heritage List identifies 1,211 properties, of which 869 are inscribed as cultural heritage, 213 as natural, and 39 as mixed sites (UNESCO, 2021a). Listing on the World Heritage List means "the universal recognition of the outstanding value of the site to humanity" and presupposes a commitment to the need for "secure a better heritage preservation" (Canale, De Simone, Di Maio, & Parenti, 2019, p. 114). Destinations use the World Heritage Site (WHS) designation as a marketing tool to improve their tourism attractiveness (Caust & Vecco, 2017; Yang, Lin, & Han, 2010) and image and competitive advantage (Canale et al., 2019; Frey & Steiner, 2011).

Destination image is a key concept in the tourism and marketing literature. Earlier studies have examined its impact on the decision-making process and choice of destination (Chi & Qu, 2008; Guerreiro, Mendes, Fortuna, & Pinto, 2020; Prayag & Ryan, 2012). Most studies examine the image of places from the perspective of tourists (Agapito, Valle, & Mendes, 2013; Guerreiro et al., 2020; Lai & Li, 2015; Martín-Santana, Beerlí-Palacio, & Nazzareno, 2017; Remoaldo, Ribeiro, Vareiro, & Santos, 2014). Few works analyze the image of a destination from the perspective of both residents and tourists (Stylidis, Shani, & Belhassen, 2017). Alternative perspectives are yet to be investigated (Nuryanti, 1996). For residents, the image of the place affects the formation and consolidation of feelings of belonging and pride (Luque-Martínez, Barrio-García, Ibánez-Zapata, & Molina, 2007). In general, the image of WHS has received little attention in the literature (Remoaldo et al., 2014).

This chapter characterizes the perceptions that public sector agents linked to tourism, culture and heritage, and the private sector (entrepreneurs and traders in tourism) have about the Historic Center of São Luís (HCSL). The specific objectives are twofold: to identify the attributes of the holistic image of the city and to detect similarities and differences between public and private sectors agents. The analysis is based on 32 semi-structured interviews with public sector heritage, culture, and tourism managers and private sector tourism and commerce entrepreneurs.

HERITAGE CITIES AS DESTINATIONS

Historic tourism cities are significant resources on which tourism develops (Ashworth & Tunbridge, 2000). The attractiveness is based on its history, architecture, the beauty of the surrounding landscape, gastronomy, artistic expression, festivals and cultural activities, and residents' hospitality (Beerli & Martin, 2004). In this way, tangible and intangible elements compose a package of attractions contributing to a city's image. The intense competitiveness to attract tourists' challenges cities to preserve the material and spiritual elements that define the image (ICOMOS, 1987); attract investment, residents, and tourists; and generate economic activities and employment.

While UNESCO focuses on mobilizing resources for conservation, the acronym WHS is widely communicated is used as one of the main engines for development (Yang & Lau, 2019; Yang et al., 2010). The literature recognizes the commercial character of the WHS brand (Buckley, 2004; Ekern, Logan, Sauge, & Sinding-Larsen, 2012; Ryan & Silvanto, 2011), used widely in fundraising for conservation and enhancement of heritage, and as an attraction with a substantial impact on the image of places (Caust & Vecco, 2017; Kim, Oh, Lee, & Lee, 2018; Su, Wall, & Xu, 2016).

Heritage Destination Image

Destination image has been a research theme in marketing and tourism since the early 1970s, beginning with the pioneering studies of Gunn (1972), Mayo (1973), and Hunt (1975). Destination image has been defined as an impression, representation, idea, imagination, knowledge, feeling, emotion, value, belief, expectation, and experience (Crompton, 1979; Lai & Li, 2015; Tapachai & Waryszak, 2000). In general, an image corresponds to a subjective interpretation of a destination with a rational and emotional nature (Lai & Li, 2015). Destination image as a global or holistic impression of a place has its roots in psychology and theories of consumer behavior (Hunt, 1975; Reilly, 1990).

The approach initially proposed by Gartner (1986) recognizes that a destination image is formed by three interrelated components: cognition, affect, and conation. The cognitive component concerns the characteristics of the place translated into attributes (Crompton, 1979) and refers to its resources or attractions (Stabler, 1995). The image of places is also influenced by feelings and emotions translated into their affective component (Agapito et al., 2013; Ward & Russell, 1981). Further, the image contains a behavioral (conative) component resulting from the influence of the cognitive and affective components (Gartner, 1993; Martín-Santana et al., 2017; Nadeau, Heslop, O'Reilly, & Luk, 2008; Pike & Ryan, 2004). The overall image component corresponds to an impression that simplifies the decision-making process (Baloglu & McLeary, 1999; Stepchenkova & Morrison, 2008; Stylidis et al., 2017). Qu, Kim, and Im (2011) integrate the two previous approaches and add the unique image component with unique attributes of the place. This approach is widely used to explain the formation of destination image (Agapito et al., 2013), including in the case of cities (Gilboa, Jaffe, Vianelli, Pastore, & Herstein, 2015; Guerreiro et al., 2020; Luque-Martínez et al., 2007), although without paying special attention to historic cities and, in particular, to World Heritage cities.

The three-dimensional approach proposed by Echtner and Ritchie (1991) is widely used to assess the image of destinations. Stepchenkova and Morrison recognize the relevance of the model by stating that,

> ...in the whole body of destination image studies, Echtner and Ritchie (1991) proposed a somewhat unique conceptualization of the destination image construct based on an extensive review of the literature on destination image research for the period 1975–1990. (2008, p. 550)

In the conception of Echtner and Ritchie, destination image should be envisioned as consisting of attribute-based and holistic components made of functional or more tangible characteristics (architecture, prices) and psychological or intangible ones (receptiveness of people, quality of service). Therefore, their approach corresponds with the cognitive-affective-overall components approach and was used to develop a scale designed to measure the holistic image of destinations (Echtner & Ritchie, 1993). This approach has been widely used to assess the image of destinations (Agapito et al., 2013; Arslanova, Agapito, & Pinto, 2017; Lien, 2015; Stepchenkova & Morrison, 2008), which testifies to its appropriateness.

Historic sites and heritage are live testimonies of the past, and expressions of local culture and identity (Fortuna, 1999). Place identity refers to the process in which identity forming attributes (locale, language, race, gender, or sexuality) serve as a limited set of raw materials for individuals to use in constructing different senses of belonging to social groups, lifestyles, and nations. Therefore, it is essential that the image projected by the management of WHS sites consider the salient attributes of the place, so that they are integrated into the definition of the competitive positioning of the brand of the city as a destination (Lewis, Kerr, & Burgess, 2019).

Study Methods

This chapter examines São Luís, Brazil, as its case study. The Portuguese implemented an urbanization project in the 18th–19th centuries inspired by the Pombaline area of Lisbon. The historic center of the city was included in the UNESCO World Heritage List in 1997 and is considered an exceptional example of a colonial town adapted to the climatic conditions of equatorial South America and where its urban fabric is integrated harmoniously with its natural surroundings (UNESCO, 2021b). The urban cultural heritage of the city is publicized as the leading local product to tourists (Mota, Guerreiro, & Pinto, 2016). Knowing stakeholders' perceptions of the destination image of the city where they live and work is essential for the development of policies in the tourism sector.

The complex nature of the destination image construct and the relevance of privileging the unique character of the place led to the choice of qualitative methods and data collection through semi-structured interviews (Bryman, 2015; Miles & Huberman, 1994). The questions asked in these interviews were derived from literature (Echtner & Ritchie, 1993; Stepchenkova & Morrison, 2008). Respondents in both groups (heritage, culture, and tourism public sector managers and private sector entrepreneurs in tourism and commerce) were selected using purposive sampling. The number of respondents was determined by the saturation criterion (Bryman, 2015). After 13 interviews were conducted with cultural and tourism heritage managers (public sector), and 19 interviews with business managers linked to commerce and tourism (private sector), the content of the information collected became repetitive, no new themes emerged, and the saturation point was reached. Interviews took place in the city of São Luís, between November and December 2017.

Data were analyzed using thematic content analysis. The content of the interviews was systematically reviewed and categorized according to its meaning (Adams, Khan, Raeside, & White, 2007; Esteves, 2006). A categorical matrix was designed for this purpose. They were structured based on the work of Echtner and Ritchie (1993) and subcategories emerged from empirical data. The information was later aggregated into units of record (discourse segments) and context (interviews) (Esteves, 2006). Each context unit was assigned a code, namely, E1, E2, ..., etc. to ensure anonymity and confidentiality. The data was imported into and organized using qualitative data analysis software (NVivo 9, QSR International). Data analysis allowed identifying the attributes of the image of the historic venter of São Luís, as perceived by the interviewees.

Study Findings

More than half of the interviewees were born in São Luís, between 36–55 years old, and hold higher education levels. The time of experience in the function comprised a minimum interval of two months and a maximum of 27 years. Of those interviewed, two respondents live in HCSL, while 17 live in other areas of the city. The sample was made up of 19 companies from the private sector linked to accommodation, travel agencies, restaurants, and handicrafts. At the level of public administration, 13 agents were linked to the tourism industry, historical/heritage, and cultural centers (Table 1).

Components of Destination Image

The attributes of a functional, psychological, and unique image of the historic center of São Luís were identified by analyzing the content of the interviewees' discourse (Table 2).

Three major categories of attributes associated with the functional image of the HCSL were identified: historical and cultural heritage, the history of the State of Maranhão, and tourism. Regarding the historical and cultural

Table 1. Respondents' Profile.

Characteristics	Sector Public	Sector Private
Gender		
Female	6	6
Male	7	13
Age		
18–35	3	4
36–55	9	11
56–66	1	3
>66		1
Education		
Basic education	–	1
High school	–	7
University level	13	11
Place of Residence		
Another part of the city	13	17
Historic center	–	2
Birthplace		
São Luis	10	11
Another city	3	7
Another country	–	1
Professional Activity		
Public Manager		
Cultural centers	6	–
Historical heritage	3	–
Tourism	4	–
Private Manager		
Handicrafts	–	3
Bars/Restaurants	–	6
Accommodation	–	6
Travel agents	–	4

heritage category, the discussion revolves around colonial architecture and urbanism, traditional, and popular culture and the state of conservation of heritage. Colonial architecture and urbanism combine in several attributes, such as colonial mansions (18), tiles (6), streets (4), urban layout (3),

Table 2. List of Attributes and Images.

Thematic	Categories	Subcategories	Subcategories
Functional characteristics of the image	Historical and cultural heritage	Architecture and colonial urbanism	Colonial Mansions Tiles Streets Urban layout Architecture Colonial buildings image from the past world heritage
		Traditional and popular culture	Culture
		Heritage conservation status	Neglect
			Abandonment and government's neglect
		History	Foundation and colonization São Luis
	Tourism	Tourists Touristic resources	
	Spirit of the place	The character of the Place	Beautiful Cultural Historic old Mystic Dirty
Psychological characteristics of the image		Feelings and emotions	Neglected Degraded Unsafe Nostalgia Frustration Enchantment Enchantment and frustration
		Hospitality	
Unique characteristics of the image	Historical and cultural heritage	Architecture and colonial urbanism	Architecture Colonial Mansions Mirantes

Table 2. *(Continued)*

Thematic	Categories	Subcategories	Subcategories
			Tiles Squares Streets Harmony of the architectural ensemble Typology of constructions
		Traditional and popular culture	Culture Reggae Tambor de crioula Bumba-meu -boi Museums Gastronomy
		History	

architecture (2), colonial buildings (2), image from the past (2), and world heritage (2). The colonial mansions are the attribute most associated with the functional component of the HCSL image. Literature referring to the architecture built in the State of Maranhão uses *casas e sobrados urbanos* (urban houses and sobrados) (Silva Filho, 1986, p. 33) to name the larger buildings. They are usually known as "colonial mansions." One of the interviewees emphasized this attribute, "the first image is the beauty of our colonial architecture with the colonial mansions that become our identity mark" (E25).

"Tile" is an attribute that characterizes the HCSL image. Starting the end of the 18th century, tiles are used in the external coating of many buildings (Silva Filho, 1986, p. 25) and other civil constructions for commercial, residential, or mixed-use. This decorative feature dates to the Portuguese presence in the region. The city became known as the "City of Tiles" whose intensive use is perceived in the comments of one of the participants, "you have a porcelain neighbourhood! Our tiles give us this strong image of the strong presence of these elements of colonial influence" (E5).

The pattern of streets in the HCSL, with the same width for both main and secondary roads (Silva Filho, 1986, p. 17), gives it homogeneity, as noted by another interview, "I also think of narrow streets. It's all very uniform" (E13). The urban layout of the historical center is an almost perfect grid, whose design is attributed to Frias de Mesquita, chief engineer in Brazil, in 1603. The orthogonal character of the urban layout is highlighted by one of the respondents, "the image I have is of the orthogonal urban layout that is our hallmark because other historic centers do not have this characteristic" (E32).

Architecture is also associated with this component of the HCSL image. "The only thing that comes to mind is the architecture and all the elements it contains" (E12). The interviewees refer to colonial buildings that reveal architectural traces from the colonial period, "I really like this kind of houses, the colonial buildings" (E11). The image from the past is associated with a specific urban context, "the image I have is that of past, past of a place, of a city" (E27).

The fact that HCSL has been classified as a "World Heritage Site" since 1997 is also associated with the image of the city's historical center, "the image of the historical center is the title it carries with it of world UNESCO heritage!" (E20). Still within the historical and cultural heritage category, but in a more immaterial perspective, the subcategory designated as traditional and popular culture emerges concerning a single attribute, "culture," which sometimes appears isolated, sometimes associated with adjectives as unique, very rich, vast, diverse, "the image is that of our popular culture, which is very rich" (E12). Only one respondent unfolds the attribute highlighting some of its elements, "the image of the historical center is that of our culture, which is handicrafts, food from Maranhão, our popular festivals, our traditions" (E18).

"Heritage conservation" status is a subcategory related to the historical and cultural heritage where attributes such as "neglected" and "abandonment," "government's neglect" are identified. Neglected appears primarily associated with a general state of lack of care that contrasts with the beauty and potential of the resources, as perceived by one respondent, "the image I have was supposed to be spectacular, but neglect of the place it's the image I have, and I'm sad about it" (E5). Still, the neglected attribute is related to the degradation of heritage due to the de-characterization of some properties that are in ruins, "the image is neglected because many colonial mansions are in ruins" (E6). Finally, neglect is associated with the omission of public authorities, neglect, and lack of incentives, "what is really happening is total neglect by the public authorities" (E16).

"History," as a category of analysis, includes the subcategory "Foundation and Colonization of São Luís" (6), referred to in a general way, "the image I have is of the history of a place" (E1). This suggests a more local and identity focus, "it is an important part of our city, of our roots, of our traditions. Anyway, the historical center preserves a little what we were and still are. It is an important space in our history" (E24).

Finally, "tourism" as category includes attributes such as tourists and the potential of the city's touristic resources, "the historical center is interesting for tourism! The historical center is unique. I associate it with a place that has a lot of tourist potential" (E11).

Psychological Image Characteristics

In the psychological-holistic component of the destination image, an attempt was made to identify the "atmosphere," "aura," or "environment" and there were three subcategories in the "spirit of the place category": "character of the place," "feelings and emotions" aroused by the HCSL, and "residents' hospitality."

Regarding the character of the place, attributes such as beautiful (6), neglected (5), degraded (5), dirty (4), cultural (3) historical (2), old (2), and mystic (2) were associated with tangible aspects of heritage. The character of the historical center was also noted, "the vast majority think our historical center is beautiful, they are impressed, they are fascinated" (E7). However, several attributes that refer to negative aspects of the neglected, degraded, and dirty environment were also highlighted.

The "neglect" attribute is directed to colonial mansions which are privately owned. Respondents consider there is a lack of political will to preserve them, "look at the image is neglect. The first impression is that we don't care that we are leaving God to give" (E16). Since most of these big houses are private, "it is very neglect by the government" (E12). The emphasis on a "degraded atmosphere" exposes the state of conservation of heritage and the question of its sustainability, "they feel that the environment is degraded on the whole. It's practically unanimous" (E8). The "dirty" attribute concerns the discomfort caused by the issue of local cleanliness and hygiene, and interviewees' embarrassment with tourists' observations, "they complain about the dirty environment" (E8).

Although less expressively, other attributes such as cultural, historic, old, and mystic were identified. The reference to a cultural atmosphere is clear, "there is a cultural atmosphere that emanates from the place with its cultural attractions" (E13). The historic attribute appears linked to architectural and urban details because "the tourist perceives a historic environment through colonial mansions, streets and construction details" (E31). The "old" attribute reveals a differentiating aspect of the place's character, "a space that is not the same as what modernity proposes today. It's an old atmosphere" (E24). The "mystic" side of the place concerns the *legends* transmitted by oral tradition, "the tourist finds a mystic atmosphere through the legends that involve a certain mystery" (E2).

Regarding the feelings and emotions and hospitality categories that make up the "Spirit of Place," attributes such as enchantment and frustration (7), nostalgia (6), enchantment (5), unsafe (5), frustration (4), and hospitability (4) were identified. Such attributes evoke the meanings of the physical and social environment, since architecture cannot be a mere game of shapes and includes feelings and emotions with the images it creates (Pallasmaa, 2011).

A mixed feeling of enchantment and frustration experienced by the tourist is the attribute that has a higher frequency of responses, "look, all this architectural heritage enchants the tourist. He is very happy to have all this in Maranhão ... But, at the same time, I see a disappointment on account of the tourist when he observes that such a beautiful place with structures inherited from most of the colonial period is so neglected" (E28). Contact with past references evokes a nostalgic atmosphere arising, which contributes to a state of satisfaction, "I think that when traveling through the historical center, the tourist feels a turn in time, a return to the past, a certain nostalgia" (E13). Enchantment is perceived as a state to be experienced by the tourist, "I think he is initially enchanted ..., the image is of enchantment" (E15).

The perception of personal insecurity is considered critical, "But there is a little security problem. The tourist becomes unsafe. Personal safety is very important when they leave home" (E11). Frustration is also mentioned, "so, he [tourist] arrives here and since our facilities, today are not ideal in the historical center, it can cause tourists a feeling of non-satisfaction in relationship with what you see, I imagine, a frustration" (E4). One of the attributes of the place is the residents' hospitality, "the tourist feels welcomed, the people from Maranhão are very hospitable. What he can find is a hospitable environment" (E1).

Unique Characteristics of Image

Regarding the distinctive attributes of the HCSL, a large category was identified: "Historical and cultural heritage," and three subcategories were created: "architecture and colonial urbanism; traditional and popular culture and history." Regarding architecture and colonial urbanism, attributes are related to the architectural component, such as the colonial mansions (11), tiles (9), architecture (5), and the "mirantes" (3). Mirantes, located on top of the buildings, are compartments aimed at environmental comfort by ensuring the internal renewal of the building's air. One of the excerpts from the interview alludes to its differential aspect in the image of the city center, "I think that the 'lookouts' (mirantes) are different elements and mark the landscape of the historical center" (E17).

The attributes most related to urbanism are the harmony of the architectural set (6), the typology of constructions (4), the squares, and the streets (2). This harmony is discussed in the response to "Criterion V" of the World Heritage List when it is revealed that "the historic center of São Luís is an excellent example of a colonial city that has preserved its urban fabric, harmoniously integrated into its natural environment, to an exceptional degree" (UNESCO, 2021a). The harmony of the architectural set is directly related to the urban layout, the division of lots into blocks and reference construction models, giving an aspect of uniformity, "Yes, this harmony of the architectural set that I have in mind" (E15). The typology of constructions is part of a wide range of buildings from Maranhão region, and it is formed by the "traditional Portuguese" style (Pombaline architecture), also by the neoclassical, the eclectic, the neo-colonial, the art deco, the modern, and the popular architecture and urbanism (IPLAM, 1998, pp. 33–41). One of the interviewees points out that "the various colonial mansions' typology that is unique, and this is different. It is remarkable here" (E24).

Despite having similar elements, each public square has a very particular ambience, "the Nauro Machado Square is what we have different due to the ambience of the place with its spaces and staircase" (E10). In HCSL, many streets are still paved in a more vernacular style, with some sidewalks paved in Portuguese limestone from Sintra (Martins, 2013), "the streets that refer to the past. This is what we remember" (E23).

Regarding traditional and popular culture, the distinctive attributes of the place emerge as the *tambor da crioula* (7), culture (6), *bumba-meu-boi* (4), *reggae* (3), gastronomy (3), and the museums (2). Culture sometimes appears in discourses in a very general way without connection to a specific expression or tradition, as evidenced, "our popular culture is very rich, there is a lot of cultural diversity (E18). It's our differential" (E18). "[P]opular culture, this cultural diversity in São Luís is unique" (E29).

However, great emphasis is placed on traditional and popular festivals and dances, with *tambor da crioula* the most referenced manifestation, followed by the *bumba-meu-boi* and *reggae*, "look, I think our parties and folk dances are unique, *bumba-meu-boi, tambor de crioula*, and even *reggae* that draws much attention. These are things you won't see in any other environment; they are peculiarities of the city" *(E27)*. It is in the functional field, "unique," according to Echtner and Ritchie (1991) that the icons and special events emerge. Museums are differentiating attributes due to the diversity of themes they add – dances, popular festivals, religiosity, gastronomy, portrait of past, and contemporary ways of life, "the museums with their different themes, I think this is unique and represents our ways of being" (E29). Gastronomy is a unique attribute due to its indigenous and African origins, "the cuisine here is very rich and typical and I realize that it only has here with all the elements that influenced it" (E8).

Even HCSL's history is mentioned as a unique attribute of the place, "Our history is what is different. The heritage is like other places in the country with few differences" (E12). The attribute history is also associated with the relationship with other peoples at the origin of the city. The French foundation is referenced as a differentiating element in the context of Brazilian colonial cities, "Our history is the French foundation that only exists here. It is our differential in the country. São Luís is the only Brazilian capital not founded by the Portuguese" (E8).

Image of Private and Public Stakeholders

A comparative analysis of the different professional sectors of the interviewees (public and private) was carried out. Regarding the functional-holistic image component, 14 attributes were identified. It should be noted that abandonment and contempt as attributes of a negative nature and which were mentioned only by the private sector. They reveal the perception of these actors regarding the conditions in which the architectural ensemble finds itself. According to Nadeau et al. (2008), images are simplifications of more complex ideas. The private sector only perceived the attributes, colonial buildings, and world heritage. The attributes that make up cultural heritage, both material and immaterial, are represented in all sectors (Figure 1).

The psychological holistic component is designed to reveal perceptions of a more intangible nature. This component aims to capture the character or quality of the environment and reveal its atmosphere or rather "atmospheres" as proposed by Zumthor (2006). Attributes related to the environment, such as

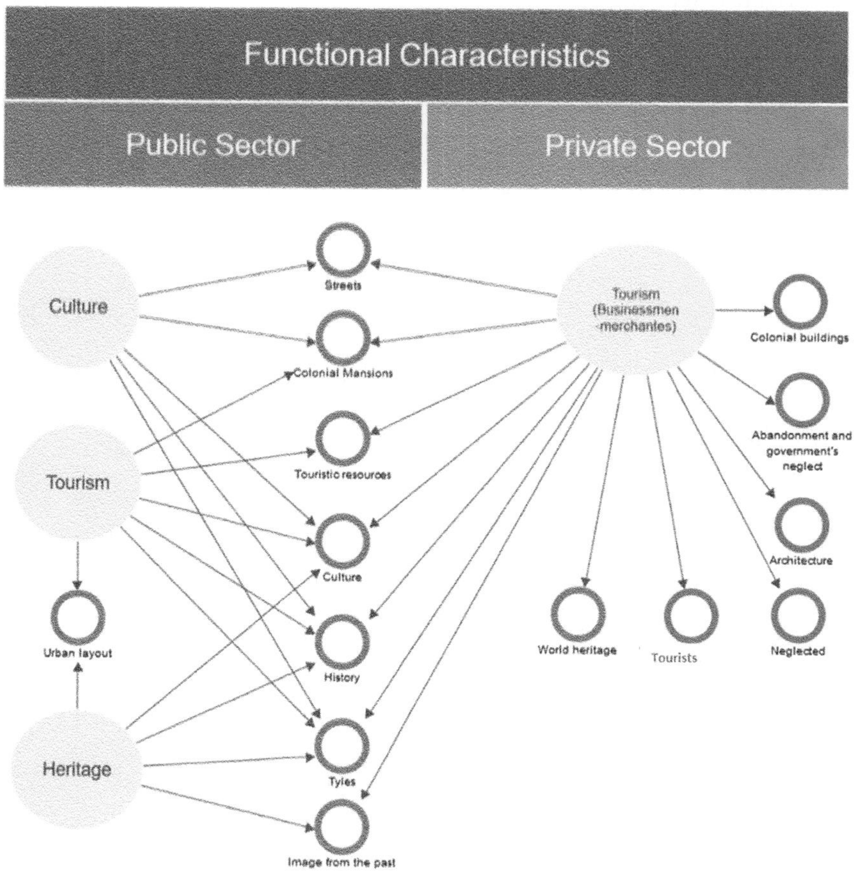

Figure 1. Comparative Analysis of the Functional Components of Image.

beautiful, cultural, historical, and local hospitability, were highlighted in earlier research (Beerli & Martin, 2004; Ferreira, 2011) and this study is referred to by respondents from both the public and private sectors. Most negative attributes are again observed by the private sector only, with an emphasis on a dirty, neglected and degraded environment. However, *A Feeling of Frustration and a Feeling of Mixed Enchantment and Frustration* are psychological attributes mentioned by interviewees from both sectors. In addition, while the focus of the heritage sector falls on a cultural and historical environment, the cultural sector mentions a cultural atmosphere, ancient and nostalgic, which are quite similar. Now, while the public sector of tourism emphasizes the ancient, mystical, and nostalgic character of the place, the private sector punctuates all other concepts except the ancient and mystical attributes. The hospitality of residents is mentioned only by the tourism sectors in the two administrative spheres, public, and private (Figure 2).

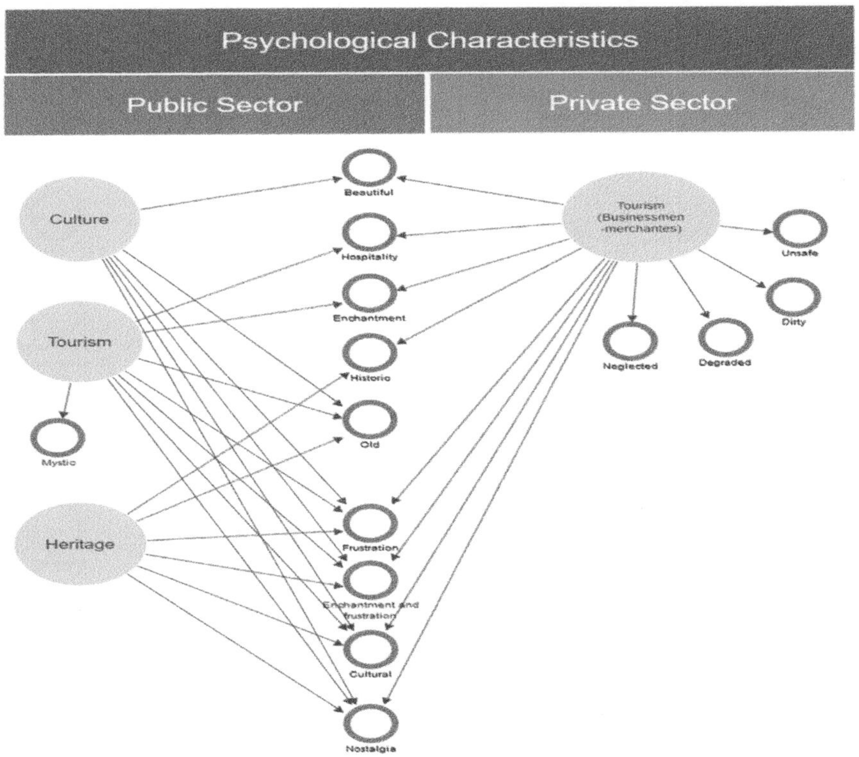

Figure 2. Comparative Analysis of the Psychological Component of Image.

Regarding the unique characteristics of the place (Figure 3), architecture and colonial mansions are mentioned by all sectors. The elements of urbanism (squares, streets, the typology of buildings, and the harmony of the architectural ensemble) are not referenced by the heritage sector, probably because such elements integrate the characteristics of other historical places, as referenced by an agent of the cultural sector, "look, the heritage here is different, it's even difficult to cite because it is so similar to other Brazilian states due to the Portuguese influence in other historical sites" (E26). Gastronomy, culture, and museums are equally represented by the private sector of tourism and culture. Cultural manifestations such as *Tambor de*

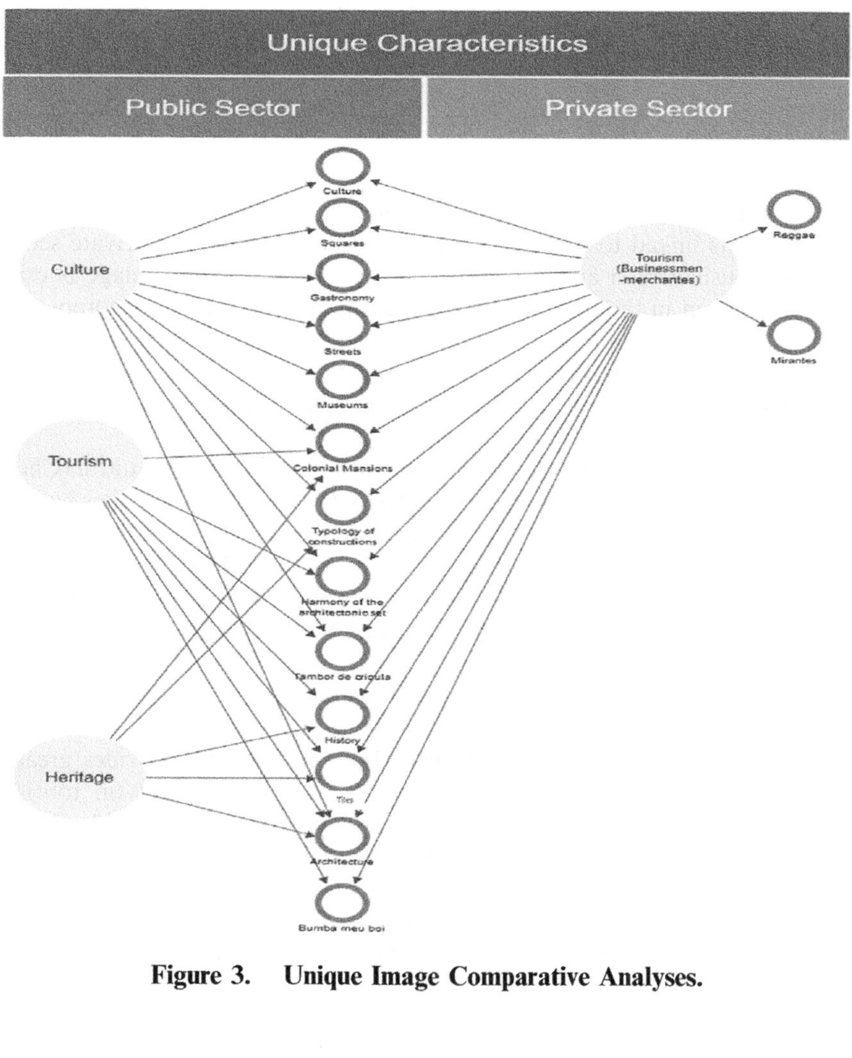

Figure 3. Unique Image Comparative Analyses.

Crioula, bumba-meu-boi, and *reggae* are not represented in the heritage sector, while the attribute of "Our History" is not represented in the cultural sector.

The study reveals that negative opinions were highlighted, mainly by respondents working in the private sector of tourism, except for all sectors' frustration attributes. This requires more outstanding care, especially when designing marketing campaigns that can generate discrepancies between representation and reality (Marine-Roig & Ferrer-Rosell, 2018).

The WHS title creates expectations of conservation and increased tourism demand and tends to be seen as a solution to problems of neglect. This often results in dissatisfaction, as shown in this interview excerpt, "the title brought very little to the heritage, it reformed some squares, some buildings, but very little. For tourism, so far, nothing. That is what you see here. There are no tourists. It generated expectations that were not fulfilled" (E12).

CONCLUSION

This research characterizes the perceptions of the HCSL image by public sector agents linked to tourism, culture, and heritage and the private sector (tourism businessmen and traders). The topic of destination image is critically significant in the tourism management process. A qualitative approach was adopted to explore the destination's image, in a holistic perspective, through semi-structured interviews. The thematic content analysis identified 35 salient attributes of the image of the city's historical center, highlighting colonial mansions, tiles, culture, and history.

Although research into the image of tourism destinations is fruitful (Chen, 2018; Pike & Kotsi, 2016), the case of UNESCO World Heritage sites has not received due attention (Chen, 2018). Most studies on this topic focus on the perspective of tourists (Belhassen & Shani, 2015; Deng, Liu, Dai, & Li, 2019; Stylidis, Belhassen, & Shani, 2015), with the perspectives of residents and local players less highlighted in the literature. Studies generally include these respondents in a joint analysis (Kong, du Cros, & Ong, 2015). The World Heritage Sites brand brings with it huge expectations (Okech, 2010), both for local populations and for those responsible for managing destinations.

Study results discern that the World Heritage List title provides greater property protection than tourism development and its effect on tourism demand is still under academic debate (Buckley, 2018; Cellini, 2011; Gao & Su, 2019). There are discrepancies in empirical studies, even in emerging destinations where a more pronounced effect is expected (Mota et al., 2016).

The characterization of the holistic image of the UNESCO World Heritage city from the perspective of local players identifies the relevant attributes of the site. This information will allow the design of an adequate marketing strategy for the destination based on local development. Furthermore, as

WHS status is used as a strategic management tool with implications for the brand of the place, the knowledge of the perceptions of key local players will allow them to position or reposition the place realistically and appropriately.

This case study has some limitations that should be seen as future research avenues. In this investigation, the perceptions of key local players with specific responsibilities in public management linked to culture, heritage, and tourism and private management with emphasis on businesspeople and traders linked to tourism were studied. Future research should also include other stakeholders, namely residents and tourists (visitors included). The interviewees' sociodemographic characteristics, which also contribute to the formation of the destination's image, were not analyzed in this case study, so future studies should explore these characteristics. Research with tourists or visitors should also consider the tourism seasonality of the place, with analyses in the high and low seasons.

Acknowledgments—This case study was financed by National Funds provided by FCT-Foundation for Science and Technology through project UIDB/04020/2020 (CinTurs) and UIDB/04647/2020 (CICS.NOVA – Interdisciplinary Center of Social Sciences).

INFRASTRUCTURE

Chapter 12

SOLAR ENERGY FOR TOURISM IN CROATIA

Diana Baus
Institute for Tourism, Croatia

Damir Krešić
Institute for Tourism, Croatia

Abstract: This chapter examines the potential of solar energy for the development of sustainable tourism in Croatia. Tourism is an important economic activity in the Croatian economy due to the mild climate and many sunny days. Solar energy photovoltaic and thermal systems can help to support sustainable tourism, as well as increase employment and cooperation between local and national governments. This study compares best practices in solar energy for the Mediterranean countries of Italy, Spain, Cyprus, and Greece. The Mediterranean Basin is a strategic development area for the European Union, and solar energy will help to maintain its stability and high-quality standards of living. **Keywords**: Solar energy systems; sustainable development; sustainable tourism; environmental protection; photovoltaics; Mediterranean Basin

INTRODUCTION

In recent years, countries around the world have begun using renewable resources to decrease pollution caused by the fossil fuel consumption. The emission of CO_2 and NO_x gases into the atmosphere cause pollution and negatively impact the environment. Furthermore, "the public has recognized that the increasing global population and growing scale of the economy cause

Managing Destinations: From Theories to Practices
Bridging Tourism Theory and Practice, Volume 14, 161–175

ISSN: 2042-1443/doi:10.1108/S2042-144320240000014012

environmental problems due to the use of the energy" (Chaoqun, 2011, p. 1722). This pollution is responsible for global warming and as a result has turned the world toward green sources of energy (Gašparović & Gašparović, 2019).

This chapter discusses current trends in electrical energy production and EU energy guidelines. These trends concern the electricity grid as well as clean energy production, environmental protection, air and sea protection, the ozone layer, reduction of carbon emissions, energy efficiency, and ambience quality through new types of lighting. Electrical energy production is important for sustainable management, given that tourism is a complex industry that implicitly affects all these factors. Tourism is a contributor to negative externalities. Clean sources of energy can partially or completely eliminate these negative effects and allow for an expansion of capacity so long as EU norms and standards for sustainable tourism are applied.

The growth of tourism in Croatia has led to environmental pollution and resource degradation. The industry worldwide has an image of being "environmentally conscious," but this is not enough to achieve targets of the Paris Agreement in 2015. For example, to meet these targets, "an average hotel would need to cut emissions by 66% by 2030 and by 90% by 2050, compared with 2010 emissions." In total, it is estimated that tourism causes around 8% of total global CO_2 emissions, with half of this an output of transportation, and the remainder related to the electrical energy used by hotels and restaurants (Becken, 2019, p. 1).

Croatia, together with 195 other countries, signed and adopted the Paris Agreement, and, therefore, is obliged to implement the climate framework. The use of renewable sources has proven to be one of the best options for protection of the environment and can provide employment and development for destinations (Gašparović & Gašparović, 2019). Croatian tourism, together with that of other Mediterranean countries, attracts 30% of worldwide volume, or around 220 million of tourists annually (Michalena & Tripanagnostopoulos, 2009, p. 667). Thus, Croatia should increase production of electricity from solar power.

During the high season (June–August), large numbers of people (tourists, one day visitors, and residents) are concentrated in Croatian destinations. This causes pressure on the public infrastructure systems, including the electrical supply network. This pressure often results in electricity shortages and even collapse of the local electrical networks. To prevent this, electricity demand must be reduced, or Croatia must import fossil fuels to produce enough electricity to satisfy the demand (Perčić & Franković, 2016). The coastal region of Croatia has up to 2,700 hours per year when solar energy can be generated (Krajačić et al., 2011), but so far, less than 1% of electrical energy (Figure 1) comes from the solar resources. The estimated technical potential of solar energy in 1% of coastal Croatia is 830 TWh/year, which is 10 times the daily primary energy consumption (Dolomez, 2012).

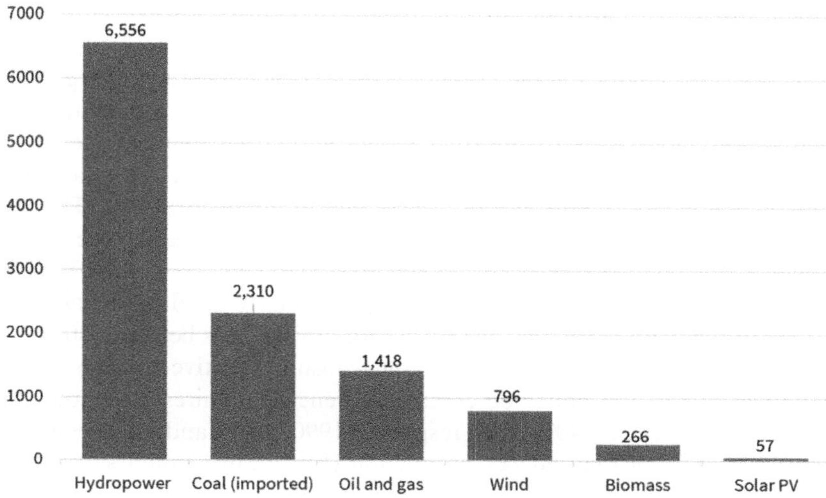

Figure 1. Electricity Generation in Croatia (GWh), 2015.
***Source:* Bankwatch Network (2019).**

Studies show that photovoltaic (PV) power plants are a good solution for inhabited islands, especially for isolated ones, because they allow energy independence and, additionally, the distribution costs of electricity are lower (Hadžić, Kozmar, & Tomić, 2014). Moreover, solar energy can be used for passive house heating, independent of the urban supply network, providing comfort with minimal environmental impact, while still retaining all the qualities and standards required in a modern home.

The use of solar energy can also reduce the negative externalities of overtourism and, at the same time, improve environmental protection and sustainable development. According to the Energy Strategy of the Republic of Croatia, renewable energy sources should by 2030 provide 60% of total electrical energy demand and 82% by 2050 (Ministry of Economy, Labor and Entrepreneurship, 2019). This will also allow Croatia to promote an image of an environmentally aware and green sustainable tourism destination.

SOLAR ENERGY IN CROATIA

The Republic of Croatia is situated in the southeast Europe and has an area of 56,594 km^2 (Gašparović & Gašparović, 2019). Croatia's favorable geographical position, mild climate, and long sunny days make it suitable for the exploitation of solar energy. In this respect, Croatia's potential is the highest of the countries in Europe based on the number of sunny hours during the year. "Adriatic islands have the highest insolation level of 2300–2800 sunshine hours per year" (Hrastnik et al., 2001, p. 505), making them ideal candidates for the collection of solar energy. Luttenberg (2015) claims that Croatia has the technical and economic abilities to generate 100% of its energy requirements from solar radiation. During January, the coldest and darkest month of the year, Croatia still has doubled the quantity of sunlight compared to Northern Europe (Runko Luttenberg, 2015). The total energy available from solar radiation based on the area of the littoral provinces (24,696 km^2) is over 125,600 PJ/a (Hrastnik et al., 2001, p. 5). South Dalmatia, Split, and Lastovo have particularly good potential.

One of the main issues that the European Union faces before 2030 is how to supply all the energy needed without causing negative impacts on the living world. The EU policy on climate and energy requires greenhouse gas emissions to reduce by 40% (with respect to 1990 levels) and green energy to increase to 32% of final energy production (European Commission, 2015). Croatia, as a member state, has an obligation to fulfill the goals set by the European Union. Croatia has developed its own national long-term strategies and "National Determined Contribution" that specify its reduction targets. In future, tourism businesses will need to monitor negative externalities, such as CO_2 emissions, and reduce them to the agreed level (Becken, 2019).

Sustainable Tourism

Tourism is one of the most important economic activities of the Croatian economy and has experienced constant growth in recent year. Over the 6 years to 2020, tourism recorded steady growth and was an important stimulant of the economy (Ivandić & Šutalo, 2018, p. 22). In 2016, some 88.34% of direct gross value added came from this industry, while other industries together contributed only 11.8%. The economic impacts of tourism, analyzed through the Tourism Satellite Account show that tourism directly contributes 11.4% of Croatian gross domestic product (GDP) and the total contribution of tourism to Croatian GDP is 16.9% (Ivandić, Marušić, Sever, Kos, & Vrančić, 2019, p. 1). Compared to other Mediterranean countries (with the exception of small island economies, such as Malta and Cyprus), tourism in Croatia has the largest share of national GDP. However, as already noted, the electrical network supply cannot keep up with demand during the peak summer months. Further, demand also exceeds supply for other resources such as water and municipal infrastructure.

The use of conventional power plants has four general disadvantages: reduction in electrical energy or lower quality of energy (drops in voltage), high environmental pollution with greenhouse gases emission, increasing costs of imported electrical energy (energy dependence of other producers), and higher costs of energy consumption. Renewable energy from wind and solar energy sources avoids these problems. However, wind power plants make noise and detract from the natural scenery. Hence, solar energy is the best choice to substitute for classical energy production, while at the same time not reducing the attraction of Croatia for tourists.

The Clean Energy for EU Islands Initiative of the European Commission (2019) targeted six Croatian islands (Cres, Vis, Korčula, Hvar, Brač, and Lošinj) with the aim to fully decarbonize the energy sector by 2040. Since February 2019, these islands have developed decarbonization policies tailored to their environment. The strategy provides complete energy autonomy for these Adriatic islands.

Positive Impacts of Solar Energy

In general, solar energy technology is not completely free from negative emissions, but it has the lowest negative impact on the environment compared to the other renewable energy sources. Solar technology can be divided into two categories, solar thermal systems and PV power generation. Solar thermal systems provide efficient and simple technology to generate hot water for households. It is suitable for locations with high solar irradiation where it can significantly contribute to reduction of energy load. Solar thermal systems have an important role in hotels, houses, camps, and many other forms of accommodation. They provide hot water during the summer months when demand is very high. Moreover, solar thermal can provide warm water for swimming pools in tourism accommodations year-round.

PV power generation produces electrical energy from photovoltage cells/panels without generating any noise. PV systems are used for "lighting for domestic and community activities such as computers, lighting, radio and phones" (Srinivasan, Velu, & Mahubabu, 2019, p. 793). Hence, it improves the quality of living and contributes to job creation (during installation, maintenance, and other activities), therefore decreasing poverty in the community. PV systems are suitable for use in any type of accommodation, or they can stand alone, and do not reduce the natural beauty of the landscape (Sánchez-Pantoja, Vidal, & Pastor, 2018).

PV systems are also used for the public lighting of highways, local streets, squares, park, traffic signals, and more. This type of lighting uses new generation lighting emission diodes (LEDs), which are low in energy consumption and have a long lifetime. LED lighting does not need to be connected to the electrical network because each has its own PV panel.

Barriers to Renewable Energy Use

The three main barriers to renewable energy use in Croatia are the national policy, energy monopoly, and social resistance. First, the energy strategy adopted in 2009 did not consider global energy trends but envisaged the construction of two conventional power plants, as well as the possibility of building of a nuclear power plant. Instead of investing in energy independence, the country used conventional energy production. Up to 2020, targets for energy development were neither transparent nor consistent. There was no:

> …vision of sustainable development as a basis for shaping public policies in the energy sector and environmental protection nor public access to credible data on potential economic, social and ecological effects of wider use of renewable energy systems. (Runko Luttenberg, 2015, p. 647)

Decision-makers and civil society (experts and organizations) were not included in the decision-making process.

Second, the national electricity company that was formed in 1990 had a monopoly on electric power production for the whole country, with production, transmission, and distribution of electricity using fossil fuels. Therefore, this company did not have interest in renewable energy systems other than hydropower. The company was intended to be the main initiator of the solar industry, but instead decided to develop a "centralized system by building the large power plants and make money by charging customers for taking power off the grid." Introduction of new energy systems would require building of thousands of miles of new transmission lines across public and private properties, so it was easier to maintain the existing investments. The Croatian government and the national electricity company decided to invest into conventional power plants, and renewable energy production was left to citizens and foreign investors (Runko Luttenberg, 2015, p. 648). Thus, the obstacles and barriers for renewable energy systems have political and technical dimensions.

Even though the Croatia public has a positive attitude to green energy, generally, there is a lack of concrete information and understanding of concepts which would be of great importance to the end users and their decision of turning toward clean energy. This lack of knowledge about renewable energy systems among citizens may be due to media reporting which focuses on the:

> …negative influence of renewable energy on electricity prices. On the other hand, people who are better informed about renewable energy, those being the ones who use internet, turn out at election in smaller percentages than individuals who are informed through newspapers or television.

Those who support PV installation or solar water heaters suffer from a lack of information, administrative barriers, and complex procedures (Runko Luttenberg, 2015, pp. 651–652).

Furthermore, colleges and educational institutions do not offer classes and formal education regarding renewable energy issues, resulting in a lack of professional and trained human resources. Additionally, local communities, due to the lack of knowledge, are not in any way stimulated to invest in sources that would be of benefit to the whole community. As a result, the government prefers large scale centralized electricity production. Croatia has not developed a new energy strategy since 2009, nor has it decided its energy strategy, which is crucial for the sustainable tourism development.

Best Practices of Mediterranean Countries

This part of the chapter examines the best practices of Mediterranean countries (Italy, Spain, Cyprus, and Greece), in using solar energy systems to achieve sustainable tourism development. The Mediterranean basin is a region of strategic importance for EU development; therefore, it must maintain its stability and high-quality standards of living. Sustainable tourism development is important in the economy of each Mediterranean country. They all have very rich natural resources, cultural heritage, and also are politically stable, which attracts mass tourism. The number of tourists to the region has increased up to 2020, especially to Mediterranean islands. In most of the cases, mass tourism often leads to environmental destruction and degradation of natural landscapes, water supply problems, pollution of coastline and islands, as well as transport emissions and infrastructure (Michalena & Tripanagnostopoulos, 2009, p. 667).

The best practices of Mediterranean countries for technology and renewable energy sources are dependent on their legislative and regulatory framework. At the national level, legislative frameworks are the subject of ministerial decisions, but at the local level they are of higher importance. Local communities, especially on the islands, have small contributions to CO_2 emissions, but are affected by climate changes at the same level as the mainland. "Moreover, the lack of electricity transport system (more than 50% of islands in Europe are not connected to the mainland), as well as the high cost of their energy demand are handicaps which should be met." In addition, supply of islands from the mainland is done through the undersea power cables that have less and less supply capacity compared to the growing demand in recent years. With the new technology, a solution is offered that energy is produced on site without additional infrastructure (connection of the island to the mainland). Meanwhile, Mediterranean islands have abundant of natural energy resources which make them ideal place for innovative renewable energy projects and the development of thematic forms of tourism. The lack of information presented to the end users in the Mediterranean area was solved with the specific strategies and practices which are based on specific motivations and several criteria (Michalena & Tripanagnostopoulos, 2009, p. 668).

The Italian Island of Sicily bases its economy on tourism. Solar energy contributes 8% of the total renewable energy, a high percentage, given that solar panels installation only began in 2010. Around 70% of electricity still comes from oil, while 25% of energy comes from renewable sources, mainly hydroelectricity (Bassio & Matera, 2011). Hydroelectricity systems are more efficient than wind plants or any other energy source.

The Greek Island of Gavdos promotes the use of solar energy to attract tourists. The peak usage is in August, when arrival numbers reach their maximum. The island is isolated from the mainland, and as it is very small (30km), so it produces its own electricity using PV units. This practice began in 1987 and continues today. Gavdos successfully achieved total decarbonization because it overcame all barriers and residents accepted the renewable energy technologies. This is one of the best examples and the inspiration for this chapter. It shows the direction for development of sustainable tourism in the coming decades.

Crete has also introduced a 5-year program to achieve a reliable energy system based on green energy sources. Among the results of this program were the introduction of PV systems in the island and the development of ecological tourism which has already offered many benefits in the island's economy (Michalena & Tripanagnostopoulos, 2009, p. 669).

Since 2008, PV panels and solar thermal collectors are widely used in Crete. This is beneficial for tourism, especially during the peak season because it assures large quantities of hot water and electricity. In this way, Crete avoids the extreme use of fossil fuels that the peak tourism season would otherwise require. The Canary Islands, Spain, contributes to the implementation of renewable energy sources by providing information to locals and decision-makers to influence them in favor of sustainable energy.

The renewable energy legislation framework in Sardinia focuses on the energy needs of separate sectors (Michalena & Tripanagnostopoulos, 2009, p. 669). Tourism in Sardinia has a major contribution to the overall economy. National authorities developed an Energy Plan for tourism because hotels and other infrastructure on the island use a large quantity of electricity, meaning it was dependent on the mainland for oil. It is decided that solar systems would be the great source of wanted quantity of electricity for the needs of hotels and tourism infrastructures.

Cyprus is the first country worldwide to use thermal collectors for the purposes of tourism but also for private purposes. In contrary to Sardinia where an energy plan was developed for each industry, Cyprus based its plan "on the local geo-morphological features, climate, local natural resources and local acceptance for innovative technologies" (Michalena & Tripanagnostopoulos, 2009, p. 670). Cyprus uses solar vehicles on luxury golf courses. For agricultural purposes, biomass is a preferred energy source.

Research on renewable energy technology for Mediterranean countries shows that wind generators have the greatest potential. At the same time, countries are reluctant to use wind technology as large turbine towers may detract from the natural beauty of the landscape. Their large size means they are not suitable for small, inhabited islands because they disturb landscape views and diminish the natural beauty. Local inhabitants report health issues because of noise from their rotating blades.

To move up to date with national sustainable action plan, many Mediterranean islands which are not connected to the national electrical network accepted and adopted solar thermal collectors and PV systems. Plenty of sunshine all year long, and long warm summer days provide consistent source of energy enough for the whole year. Furthermore, thermal collectors are more cost-efficient way than other renewable energy sources (Michalena & Tripanagnostopoulos, 2009, p. 671). Other sources of renewable energy ask for the high subsidies from the national governments for their implementation (such as implementation of wind power turbines, hydro power plants, and the like) because their start-up is a question of a national budget, not a single-person decision. Therefore, solar energy systems offer a great level of independence, from the fact that one can choose it as their electrical supply preference, to the fact they can choose how and where it will be installed (standalone device, roof panels, ground panels, etc.).

The use of electricity can be divided for heating and cooling needs, as well as for the electric supply for the rest of the household appliances. "Solar energy can cover a significant percentage of tourism infrastructures, like hotels energy needs (hot water and space heating during winter and air conditioning during summer)." PV devices can be applied in:

> ...other sectors besides tourism buildings (like for example in the agricultural sector, such as greenhouses and product drying), or can also be used for water pumping, water cleaning and disinfection, and water desalination which is important since water scarcity is one of the most important problems of islands. (Michalena & Tripanagnostopoulos, 2009, p. 669)

Research shows that national policy frameworks in many Mediterranean countries have not encouraged large scale use of solar energy systems as conventional power suppliers have a monopoly and may see renewables as competition. To have an effective energy consumption, energy savings come first in all cases with wide application of solar thermal collectors, and after that, wind turbines and PVs could be applied. The only country which serves as an example of wide solar energy implementation for domestic and public need is Cyprus. The rest of the countries still wait for the "green light" and initial investments to be done by the national and local governments.

Since 2009, 30% of population in Greece (mainland and islands together) have installed sun energy systems to heat the water. The boost of solar thermophonic flat collectors happens as the result of their flatness and esthetics. "In order to have an effective energy consumption, energy savings come first in all cases with wide application of solar thermal collectors and after that, wind turbines and photovoltaics could be applied" (Bassio & Matera, 2011, p. 3).

The comparison of good practices already realized by the Mediterranean countries provides an incentive to apply and develop the same trends in Croatia. Realized practical examples undoubtedly indicate that the positive trends in the development of sustainable tourism could be applied and even better developed in Croatia due to its geographical benefits. The first PV power plant was put in operation in Croatia in 2020. It is located on the farthest inhabited island, Vis, and provides 3.5 MWa, sufficient to supply the 1,600 households on the island. Croatia plans to build eight other PV power plants on other locations along the Adriatic coast.

Study Results

The results obtained from the Mediterranean islands were studied based on their similar characteristics to identify their best practices. The results indicated that solar energy is a type of renewable source with a significant role in ensuring prosperous sustainable tourism development. At the same time, results can be affected by barriers caused by the political and strategic characteristics of the place, as well as challenges in achieving an understanding of renewable energy technologies.

The barriers mentioned earlier affect 10 million of the residents in the Mediterranean countries who face the lack of adequate access of electricity and, also, cope with urgent socioeconomic problems. Knowing that Mediterranean islands contribute to the national GDP, they are also "regional economic–political systems" therefore, their governance body could influence the expansion policy for renewable energy systems (Michalena & Tripanagnostopoulos, 2009, p. 670). This imposes the need for islands to create a strong "political power of influence." This will create comparative advantage, as well as added value (Michalena & Tripanagnostopoulos, 2009, p. 670). All mentioned case studies are subject to political and strategic decision for the purpose of realization of renewable energy systems as well as to promote sustainable tourism development.

The potential of mass tourism depends on the activities of governments of Mediterranean countries, which together with the local inhabitants could make a specific model tailored according to unique needs of a certain place.

> Within this scope, 'threats' identified in local natural, economic and social contexts could give birth to 'opportunities' regarding renewable energy systems implementation and could turn to comparative

> advantages for islands through autonomous and sustainable energy infrastructures. (Michalena & Tripanagnostopoulos, 2009, p. 670)

Sicily faced a high unemployment rate which was reduced with the development of new technologies in energy that led to a new thematic type of tourism. In comparison to Sicily, Gavdos' policy concentrated on the implementation of PV energy systems on house and for the purification of water for the purposes of drinking and agriculture. The application of renewable energy system will differ from island to island based on the type of tourism and activities that the island wants to develop.

> The understanding of the relation between renewable energy systems and tourism development is essential, since they could constitute to a leverage and motivation for the development of specific tourism activities. (Michalena & Tripanagnostopoulos, 2009, p. 670)

Meanwhile, the potential of implementation of solar energy on the islands that have not been exposed to mass tourism is an opportunity for further development of the tourism in a sustainable manner, but also it gives an incentive and a good practice example to increase the growth of the local economy. The result of that practice would be creation of more sustainable tourism, which is of great interest to undeveloped islands. In contrast to small, energy dependent islands, larger islands are focused on other industries besides tourism. Therefore, in some cases, the model of renewable energy systems cannot be chosen.

In Mediterranean countries, such as Cyprus and Canary Islands of Spain, the development of tourism is under the control of private interests and investments. In these cases, motivation for renewable energy systems is not dictated by institutional frames but mostly by the private interests. Sometimes, that leads to neglect of the local needs and requirements for environmental protection. In contrast to Cyprus and Canary Islands, Sardinian tourism development depends on national strategy framework where development is framed by the national institutions. The focus is on the development of local communities with the emphasize on building local skills and technological innovations. These strategies for development of sustainable tourism are funded by the EU fund, but also receive funding from the international institutions.

Financing of different projects regarding solar energy applications is provided depending on the existing tourism needs and chosen national policies. The process of making actions and finding methodologies for the implementation of solar energy systems opens the space for the exchange of opinion of all policymakers and local actors. The focus is on the stronger collaboration and open dialogue with the target of effectiveness through finding local entrepreneurs and other relevant institutions. The end goal of all stakeholders in the project is to find the best solution on promotion and achievement for sustainable development of tourism.

> Especially for the Mediterranean islands, strategies should aim at preserving nature and heritage, developing society in a sustainable way, building energy capacity and energy autonomy, and achieving economic growth. (Michalena & Tripanagnostopoulos, 2009, p. 670)

Local governance has the main role for solar energy system to be adopted by the local inhabitants. Their role is to assure that local community accepts the concepts of innovative technologies, as well as to ensure that they participate in the process of choosing this system. Sometimes, it requires time for these two actions to be accomplished. For the successful implementation of the technologies, local communities must be informed about the details and benefits, but also about costs that new systems could expose them to. Research shows that the acceptance level depends on previous positive experiences; therefore, the priority for the future projects for implementation of solar energy systems should be given to those who already have experience, and who are ready to accept even higher level of renewable energy technologies.

> Within the procedure of identifying the appropriate approach, local governance should be powerful, stable in its decisions and initially independent from the central national one to succeed a possible synergy between a "top-down" and "bottom up" approaches. (Michalena & Tripanagnostopoulos, 2009, p. 670)

Cooperation on the level of national initiatives and local instruments requires synergy with the aim of achieving targeted changes for the acceptance of new innovative technologies. In the Mediterranean basin, underdeveloped areas can be supported by those countries that have already had experience and that have successfully implemented renewable energy sources in their policies. In that way, undeveloped countries have a possibility to establish the best solution, including energy mix and practices for specific regions and uses.

Implementation of Solar Energy Use

In Croatia, there is enough interest for investments in solar energy systems. On the local level, there is an opportunity for citizens to participate in the procedure of decision-making and can choose the best option for the energy supply in their community. In case of increasing efforts in the field of renewable energy systems with the focus on solar energy, there are several advantages that would contribute to the growth of domestic economy in the long term, such as job creation and energy revenue reinvestment in the local communities. The investments in renewable energy sources are presenting an important role to become independent from the conventional energy sources.

The responsibility for energy-related emissions and other pollution falls on cities and local governance. They could spread and promote the concept of solar energy use for electricity production and heating. Hence, municipality authorities could use their set of tools, such as "building regulation for new constructions and reconstructions, and through its municipal utility company." The island of Krk is a good domestic example and will be the first island in the Mediterranean basin which will adopt complete decarbonization by 2030. The report "Interdisciplinary greenhouse gases zero-emission strategy for integrated sustainable development" (Runko Luttenberg, 2015, p. 653) suggests that Croatian governments at all levels and public sector utilities must oblige all national institutions to use solar systems on their roofs. Also, to everyone else, they should give a choice on the mode of power that will be protected by the national legislative and European directive (2014/25/EU/59).

Croatia, as a Mediterranean tourism destination, should follow global trends with regards to architecture and new technology. Passive house is a new approach for the construction of accommodation capacities (private residential houses, hotels, bungalows, etc.), with a very high ergonomic comfort level, but at the same time, with the minimal or zero consumption of electrical energy from the public network supply. Passive solar architectures are constructed in a way to have a double role, to be solar collectors but also to act as a heat storage tanks. In general, passive houses are not more expansive comparing to the classic houses, but the difference comes from the functionality and smart design, not in a special technology. The advantage of passive solar house is in low costs for maintaining, as it does not require any additional equipment as well as additional maintenance.

The construction of the passive house is specific because it has windows with three thick panes, very thick thermal insulation, and controlled ventilation. All three elements ensure that inside space makes pleasurable living conditions, regardless of the seasons. Balanced temperature is achieved by the accumulation of the collected heat during the day; therefore, the same heat can be used over the night. Hence, there is no need for conventional sources of heat. The purpose for building passive houses is to enable consumers to produce their own energy and, in some way, to be energy independent. Regarding the promotion of passive solar houses, Croatian legislation supports such buildings with co-financing up to 80%. Encouraging financial incentives and conditions would enable new types of tourist accommodation capacities on the coast.

This new approach in self-supplying energy buildings (they produce electrical energy and heat the water with the solar energy collected from the solar panels on the house/roof) is ideal for Croatian islands. "Islands like Hvar, Vis, Korčula or Lastovo have more than 2,700 solar hours per year"

(Hrastnik et al., 2001, p. 5). Therefore, this type of constructions could be ideal for summer tourism, but also would give a boost to slow winter period. Winter tourism is low due to the colder days and inadequate and expensive solution for heating tourist accommodations. However, until today, Croatia put the emphasis on such types of buildings, but it still relies on the traditional, old-fashion style in construction (Veršić, Muraj, & Binički, 2018). Nevertheless, building of the passive houses that encompass efficiency in zero-energy consumption presents a new trend in architecture for the future generations and upcoming tourism strategies regarding sustainable development.

CONCLUSION

Croatia, with its coastal areas and islands within the Mediterranean basin, is rich in natural resources and cultural heritage which plays an important image of a "cradle of civilization" in Europe (Hrastnik et al., 2001, p. 1). This area, with mild climate, famous for its untouched nature and clean maritime zone, offers unique chance for the development of tourism. Besides the opportunities in tourism, climate gives space for the potential development in the field of energetics, particularly renewable solar energy, in the great measure. Within this perspective, solar energy systems constitute the main devices to meet energy demands in tourism and other sectors of national industries. The construction of solar power plants in Croatia would ensure sufficient energy needed to meet all requirements to achieve successful and sustainable tourism through the whole year.

Solar system gives the full set of socioeconomic advantages (Srinivasan et al., 2019). It improves the quality of living and reduces migration, creates news job opportunities, and reduces social inequalities of the community. Hence, it also impacts the national economic growth in a long-term period. Moreover, it is an alternative to mitigate the impacts of negative outcomes of overcrowding from excessive tourism and better spatial planning of the coastal area.

From the social aspect regarding the visual impact, solar energy devices can have a visual intrusion and be perceived as unattractive. The visual impact usually highly depends on the frame design and area where it is situated. However, due to the increasing orientation toward the technology development on the energy-saving and environmental protection, the global market of the solar panels is quite developed so that each year it is possible to find smaller, sampler, and more efficient solar panels.

Reduction of the CO_2 emissions, as well as small degradation of water quality supplies, put the shadow on the negative visual impacts of solar energy installations. Nowadays, people show much more concern about the ill effects of pollution than care about esthetical aspects. This demonstrates the fact that a local state of mind shifts in favor of renewable energy source acceptance, regardless of high initial costs (installation costs and services) comparing to the conventional economic standards. The public' interest in more investment in solar energy increases proportionally with an increase in knowledge about renewable energy systems. However, the vast majority of Croatian public is still insufficiently informed.

State authorities should introduce legislation to enable the use of renewable energy, particularly solar energy generation. This is achievable with the collaboration with the main electricity supplying producer company, public institutions on national and municipality levels, public entrepreneurs, and stakeholders as producers but at the same time consumers of electrical energy. Local communities have the responsibilities to implement the task for solarization of their own local territories.

> They have inherent autonomy to regulate the situation on site, particularly via their well-established municipal water, wastewater, and solid waste utilities companies which may provide other renewable energy systems as well for producing the energy locally (biogas from biowaste, and sewage sludge, energy accumulated through water lifting, own roofs and other appropriate surfaces). (Runko Luttenberg, 2015, p. 654)

Croatia and other Mediterranean countries are encouraged to follow the EU directives regarding the decarbonization and environmentally friendly technologies like renewable energy systems and energy efficiency. International financial institutions are ready to support innovative projects and their operative phases.

This eco-friendly environment is recognized by the Croatian high-quality technical colleges and academic institutions that participate in the education of young professionals with the aim of readiness to respond the challenges of new technologies like rentable energy generation, such as solar energy systems. The challenges are also reflecting on the education of lower levels, like in middle and even elementary schools. Moreover, it involves all participators like teachers of young citizens and other users of environmental utility service to be ecological conscious (Runko Luttenberg, 2015, p. 654). To remove the false message, special attention should be given to the journalism and educate the journalist to spread the positive and reliable information about impacts and outputs of solar energy systems.

Chapter 13

ARBITRATION IN TOURISM
The Industry Performance

João Vidal
University of Algarve, Portugal

João Albino Silva
University of Algarve, Portugal

Guilherme Castela
University of Algarve, Portugal

Abstract: This chapter addresses the issue of arbitration in tourism from the perspective of litigation. International tourism requires two or more legal systems to solve a given problem, which creates great complexity. To diminish this effect, organizations have been encouraged to use arbitration instead of courts, and thus, it is necessary to measure its effect on them. The analysis used the Biplot methodology, a multivariate technique in the context of reduced dimensionality. The results obtained indicated that tourism demand and supply were willing to pay for arbitration, but the latter erroneously considered the former unwilling to pay. This chapter suggests that tourism companies can raise the price of their products to increase their profit. **Keywords**: Dispute; arbitration; performance; competitive advantage; Biplot; tourism

INTRODUCTION

International tourism deals with different legal systems more than many other activities. In a dispute situation, two or more legal systems may effectively be called upon to solve the problem, with consequent articulation

Managing Destinations: From Theories to Practices
Bridging Tourism Theory and Practice, Volume 14, 177–191

ISSN: 2042-1443/doi:10.1108/S2042-144320240000014013

problems. Insurmountable cases often deny reimbursement for their damages to injured individuals.

To overcome this problem, the use of alternative dispute resolution processes (Böhmelt, 2010) has increased and been encouraged. Among these processes, arbitration is of particular importance (Alpa, 2003; Benedettelli, 1997; Beraldo, 2014). Arbitration involves a set of arbitrators performing as judges of a particular case who set out all the steps deemed appropriate for a fair dispute resolution. Other different processes such as mediation are available. However, in arbitration, the arbitrators can decide and impose their decision like a national court, while in mediation, the mediator has no power over the parties to impose a decision. Instead a mediator tries to obtain a solution in which parties can both agree on.

There is increased need for the implementation of instruments of international dispute resolutions, while these are still the subject of discussion (Zelst, 2018). Although their use in public law has already been agreed (Villa-Lobos & Vieira, 2010), arbitration has been particularly significant in the private sphere, especially regarding international commercial law (Bühring-Uhle, 1996; Cordeiro, 2015).

The main goal of this chapter is to assess the effect of alternative dispute methods on the performance of the tourism companies. This assessment clarifies whether there is a competitive advantage for companies to adopt this type of dispute resolution mechanism. This will determine whether, in equal situations, in which the only variable was the use of arbitration, this differentiation brings a real benefit to organizations Differentiation appears as a label for a wide range of strategies (Mintzberg, 1988) and allows something to be created that can be considered unique in the sector in which the organization operates. This is to further determine whether, albeit at a marginally higher cost, the institutionalized adoption of arbitration benefits organizations.

Arbitration

Arbitration generally consists of a litigation process characterized by the attribution of jurisdiction to one or more persons, chosen by the parties themselves or by third parties, whose decisions have the same efficacy as court judgments (Cordeiro, 2015). Therefore, arbitration is "a means of dispute resolution between two or more parties, effected by one or more persons who have powers to that effect recognized by law, but assigned by agreement of the parties" (Barrocas, 2010, pp. 31–32).

An arbitration convention is the agreement of the parties to submit the resolution of one or more determinate or determinable disputes to arbitration, in line with the concept adopted by the New York Convention on the Recognition and Enforcement of Foreign Arbitral Awards of 1958 in Article 2, Paragraph 1. Therefore, its object may be a current dispute (arbitration commitment) or possible disputes arising from a particular contractual or extra-contractual legal relationship (commitment clause) (Barrocas, 2010; Cordeiro, 2015; Mimoso, 2009; Pinheiro, 2005).

Internal arbitration is performed in a particular legal order, usually coincident with the state/nation. Directly and in a regulated manner, arbitration replaces the courts of that state and promotes speedy decisions. On the other hand, international arbitration, or transnational arbitration, as some authors prefer to call it (Pinheiro, 2005), brings into play interests of international trade or, in another formulation, interests of different national legal orders. In international trade relationships, arbitration is now a fundamental means of dispute resolution. It is estimated that about 90% of international trade contracts contain arbitration agreements (Pinheiro, 2005), and in some sectors of activity, such as civil aviation, it is even the normal way to resolve these disputes (Vicente, 2004).

In addition to its legal aspects, arbitration is also, and above all, a socioeconomic phenomenon. It is sociological because the social fabric demands more and better processes of dispute resolution, to which the legislators are sensitive, and economic because the field of arbitration development cannot but be the economy itself and, within it, national and international trade. Cordeiro (2015) considers that the enormous influence of arbitration on international trade indicates this procedure is a pioneer in globalization.

A Factor of Business Competitiveness

Competitiveness is a concept based on the comparison of the capacities and performance of a given company with those of another entity (Garelli, 1997), a sector of activity with another, a destination with another, or even a country with another (Krugman, 1997). On the other hand, the concept of performance is based on the comparison between the results expected and those achieved (Rumelt, 1984). Performance concerns one organization and is unrelated to other ones. This comparison will necessarily require the presence of processes to monitor and measure these results (Lucena, 1977).

The word "strategy" may have several meanings, depending on the context in which it is used (Tribe, 2010). The most common and frequent meanings have to do with policies, objectives, tactics, goals, or programs. An effort of conceptualization will be required to find the correct definition within the scope of interest (Mintzberg & Quinn, 2001), but it is always regarded as a measure of economic success (Ritchie & Crouch, 1993). The criteria that make a company more competitive in a given market may vary, given that it may hold a significant percentage of the market, a particular share in certain niches, or impose upon competitors through prices, one of the most classic and operative factors of competitiveness.

> Ritchie and Crouch addressed the concept of competitiveness, concluding that Superficially, competitiveness seems to be a simple concept about which there is little disagreement... Therefore, it is when we try to measure competitiveness that we begin to understand the difficulty in defining it, because competitiveness is a relative concept and usually multidimensional. (2003, p. 12)

Such problems led to several concepts of competitiveness appearing in the literature precisely to define the concept in use, especially in the field of tourism. Various concepts of competitiveness focus on the results of a given performance within a competitive economy frame rather than with the causes of these results. Ritchie and Crouch concluded that "There seems to be no generally accepted definition of competitiveness. It is perhaps a too broad and complex concept, defying attempts to encapsulate it in universally applicable terms" (2003, p. 13). To overcome these difficulties, they proposed a model of competitiveness analysis in order to understand what lies behind it (competition). In this model, they identified several elements and levels of competition that, when observed in an integrated way, allowed an understanding of the problem of competitiveness.

The adherence to arbitration with the purpose of gaining competitive advantage leads in a clear and logical-deductive way to "generic strategies" (Mintzberg, 1988) and, within them, to the strategy of differentiation (Sharp & Dawes, 2001). Submitting any dispute (which may arise in a particular contractual relationship between two or more parties in the field of tourism) to arbitration will never represent a classic strategy of product creation or prospection of new markets.

Ansoff (1965) used his matrix of four strategies to establish important and decisive concepts and approaches in the field of business strategy. However, although decisive for the emergence of strategic concepts (Mintzberg, 1973, 1978), this model is not comprehensive and leaves out important aspects of strategy design. It incorporates a vast set of strategies (which needed to be autonomous and conceptualized in order to give them greater and more effective operability) into the so-called "product development."

Aware of this problem, Porter (1980, 1985, 1998) presented "generic strategies," as already noted, namely leadership based on cost, differentiation, and focus. Porter considers the heart of the problem lies in identifying the strategy, criticizing Ansoff's approach for focusing only on expanding existing strategies. As can be seen, Porter's generic strategies explicitly include differentiation as a strategy for obtaining competitive advantage. In this strategy, the company invests in its image, technology, technical assistance, distribution, research and development, human resources, and market research in order to create a differentiating effect that can capture consumer attention and, in a second phase, make these customers loyal to the company.

Mintzberg (1988) was aware of this evolution, and also of the difficulties and weaknesses, and presented his generic strategies which, apparently, were intended to make the synthesis of Ansoff and Porter's thoughts, proposing new variables and recombining them according to a somewhat different logic (Ansoff, 1991). Mintzberg (1988) discussed the meaning of the word "generic" (when qualifying a strategy), recognizing that most authors used the word in the sense of any identifiable common strategy (Porter, 1980, 1985, 1998), with the sense of global strategies (strategies integrated in clusters and unrelated to individual aspects), and he himself assumed that a singular strategy could be "generic." Mintzberg (1988) focused on strategies of differentiation, establishing that basically all strategies involved a greater or lesser degree of differentiation.

Price continues to be the most basic differentiation strategy. This discussion is mainly concerned with the so-called "differentiation based on support." Mintzberg (1988) stated that differentiation based on support (or peripheral, as the author suggests that it can also be called) was more substantial than differentiation based on price or the projected image of the product, and added that this differentiation had no effect on the product itself. In other words, the product will be more attractive (or at least intended to be) with the addition of something that – without interfering with the essence of the product – gives it a distinctive ability.

The whole strategy is aimed at obtaining competitive advantage; otherwise, it would be meaningless (Reed & DeFillipi, 1990). Competitive advantage can be defined as economic advantage obtained in a contested market (Weiermair, 2000). In other words, it will generally correspond to a difference between two competitors in any conceivable dimension, allowing one to create more value than the other (Ma, 2000). Competitive advantage has traditionally been closely associated with companies.

A different and possibly more complex issue is measuring competitive advantage. In fact, this measurement, which allows organizations to be compared, is often difficult to obtain. Rumelt (2003), among others, recognized the existence of a certain convergence in the concept of value creation, a concept that is not exempt from doubts and multiple interpretations. In fact, value creation is a broad concept. It can be seen from a number of perspectives, such as that of customers, employees, suppliers, shareholders, or other stakeholders of the company, necessarily including several objectives according to their respective sensitivities (Lepak, Smith, & Taylor, 2007).

The process of competitive advantage measurement has been successively addressed; Barney (1986), Calil (2006), and Barney and Hesterly (2007) suggest methodologies of quantitative/qualitative competitive advantage measurement to evaluate the economic and financial performance of organizations. In the context of this discussion, although there are multiple concepts and approaches to define competitive advantage, the proposal put forward by Peteraf and Barney (2003) is followed here. These scholars affirm that competitive advantage is not restricted to the best competitor in a given market but to all those who create more value than the competitor that operates immediately above the break-even point.

The purpose of this chapter is to determine the factors and respective underlying values of competitiveness occurring in the phenomenon of voluntary submission of companies to an institutionalized system of international arbitration. It assesses how competitive advantage can be visualized and how it can influence the adoption of organizational strategies in the search for competitive positioning. Companies with competitive advantage would then be able to exploit the asymmetry between the maximum willingness to pay on the part of their customers and the average price charged by the market. This way, they will charge a marginally higher price, which will necessarily result in higher profit margins (Porter, 1985).

Conceptual Model and Research Hypotheses

The literature review suggested to a model of arbitration in tourism (Figure 1). Based on this conceptual model, a number of hypotheses emerged. They were tested in the empirical research, as described below. The starting point on this path can be formulated by several questions. One, considering identical situations, in which the only variation is the use of international arbitration by the tourism company, does this fact bring benefits to organizations? Two, does the

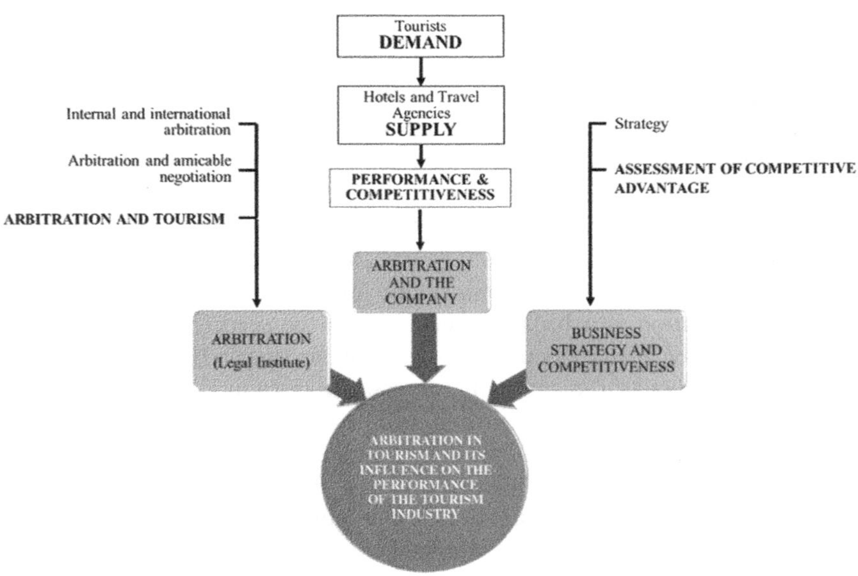

Figure 1. Model of Arbitration in Tourism.

institutionalized adoption of arbitration continue to bring benefits to organizations, even with marginally higher costs of the goods and services that they make available to customers?

Associated with the goals of this discussion, the two questions led to three hypotheses with five objectives: to identify the role of arbitration in the general framework of dispute resolution, to ascertain the knowledge of agents of the tourism market about arbitration mechanisms, to establish a litigation-based relationship between arbitration and the tourism company, to characterize the tourism market and identify the relationships between its stakeholders, and to establish the relationships between tourism demand and supply chains susceptible to generating litigation. These objectives contributed to the formulation of hypothesis *H1*:

H1. It makes sense to use an institutional system of international arbitration in the tourism industry.

The specific objectives "to determine whether or not the submission to an institutionalized scheme of arbitration promotes a better performance of tourism companies" and "evaluate and measure competitive advantage that results from the adhesion of tourism companies to arbitration schemes" contributed to the formulation of hypothesis *H2*:

H2. In terms of performance and competitiveness, tourism organizations benefit from voluntary adherence to institutional schemes of international arbitration.

Finally, the specific objectives "to determine the willingness of tourism market agents to pay for arbitration processes" and "to determine the perception of supply chains about the willingness of the demand to pay for arbitration" contributed to the formulation of hypothesis *H3*:

H3. Although tourism organizations have to market their goods/services at a marginally higher cost because they have adhered to an international arbitration scheme, they continue to benefit from better performance and competitiveness.

ARBITRATION IN TOURISM

The first goal of the study was to test a key question, which sought to confirm, or to deny, the existence of competitive advantage for tourism companies that adhered to an institutionalized system of arbitration in their commercial relationships, especially in cases when commercial relationships are international and bring different legal systems into confrontation. The existence of this competitive advantage, if confirmed, would result in better performance of the company in the market in which it is inserted.

This aim, coupled with the decision to collect primary data at the source, whether related to the tourism supply chain or to demand, in order to relate them and increase the explanatory power of the study, led to the choice of a quantitative method based on a statistical technique of data analysis. Data collection was carried out using questionnaires (Ghiglione & Matalon, 1992) applied to top managers of tour operators, travel agencies, 4- and 5-star hotels, and customers/tourists in the Algarve region, Portugal.

There are several techniques of multivariate data analysis used in these investigations, with well-defined objectives in the field of tourism. In fact, most of these techniques fall within the scope of data classification, dimensional reduction, and, above all, the detection of patterns of interest in databases using graphic factorial methods. Generally, graphical methods for factorial analyses present the results in the form of scatter diagrams, usually in a two-dimensional subspace, although the original configuration might be larger in size.

Among the methods of multivariate data analysis, Biplot methods have been a source of continuous contributions to science since their inception by Gabriel in 1971. However, these methods are still little known in the field of tourism. A Biplot is a graphical representation of multivariate data. Just as a scatter diagram describes the joint distribution of two variables, a Biplot represents three or more variables (Gabriel & Odoroff, 1990). It approximates the distribution of a multivariate sample to a small dimension space (usually two-dimensional) and overlies over the same representation, the variables used to measure the sample (Gower & Hand, 1996). The representations of the variables are usually vectors, and coincide with the directions that best describe the individual changes of each variable.

In 1986, Galindo proposed what he called HJ-Biplot. This method is suitable for the characteristics described in this chapter. In this particular representation, the interpretation of the individuals and the variables is the same. However, this procedure does not hinder the search for the variables that determine the differences between the individuals through factorial axes. This way, the method allows the interpretation of new variables, linear combinations from onset, and their relationships with the variables observed (Galindo & Cuadras, 1986).

Methodological Scheme

The empirical study reported in this chapter was conducted using three surveys. The first assessed tourists, the second assessed 4- and 5-star hotels in the Algarve, and the third assessed tourism operators and travel agencies, also from the Algarve. The first survey was applied in October 2013 and the second and third between March 2015 and January 2016. The survey resulted in 200 valid responses from tourists, 12 from operators and agencies, and 57 from for hotels. The methodological process was based on seven phases, as shown in Figure 2.

The data collection to study demand took place in the departure area of Faro International Airport, Portugal, between 27th October and 2nd November 2013. The objective of the study required use of nonprobabilistic convenience sample of tourists (Cochran, 1965; Hansen, Hurwitz, & Madow, 1953; Sudman, 1966). The sample was based on data related to arrivals at the airport by nationality, which had occurred during the same months of data collection in the previous 2 years. Unsurprisingly, it was found that the United Kingdom was the country of origin of an overwhelming majority of passengers, followed by Germany, the Netherlands, and Ireland.

A questionnaire instrument was used due to its objectivity and the ability to obtain answers in a structured way (Ghiglione & Matalon, 1992; Quivy & Campenhoudt, 1992). To help design it, some pilot interviews were carried out, either with top managers of operators and agencies or with tourists, to assess the best formulation of the questions. As they were based on scenarios given to the respondent, it was critical that the respondent understood what was being asked, and thus, face-to-face interviews were conducted (Sampieri, Collado, Lucio, Murad, & Garcia, 2006). The questionnaires were designed to characterize the respondent and to obtain answers to a set of questions about the object of study, such as "Please imagine that you have a problem related to your trip (flight, accommodation or any other facilities) that demands an application in court to be sorted out. What court would you choose based on the following criteria?"

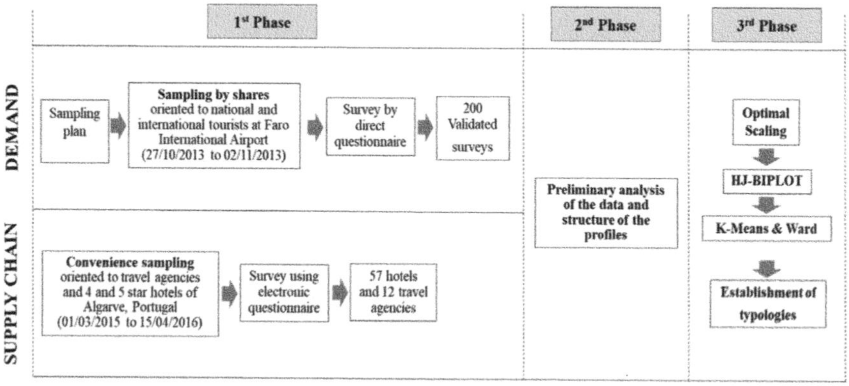

Figure 2. Methodological Scheme.

Study Results

The factorial structure of the HJ-Biplot (Vicente-Villardón, 2011) obtained for the demand is illustrated in Figure 3. The most relevant information about the active variables under study is shown (in blue) in the first axis: court, competitiveness, time, cost, and willingness to pay. The second axis has the relevant information about the active variables: problem and heard of.

By association of the five active variables with the first axis, it was possible to name it "willingness to pay for the arbitration process," which represents 66.9% of the original information retained. On the other hand, the association of the variables "problem" and "heard of" with the second axis led to name it "perception of the arbitration process" which represents 17% of the original information retained. The segmentation of the individuals into three clusters was obtained by means of cluster analysis according to the k-means nonhierarchical method and is described in the same factorial structure of the HJ-Biplot (Figure 4).

The association of the four active variables with the first axis led to name it "perception of the arbitration process," which represents 35.3% of the original information retained. Likewise, the association of the variables "court," "territorial competitiveness," "time," and "cost" with the second axis led us to name it "perception of the arbitration process," which represents 34.2% of the original information retained. The segmentation of hotels and travel agencies into three clusters was obtained by means of hierarchical

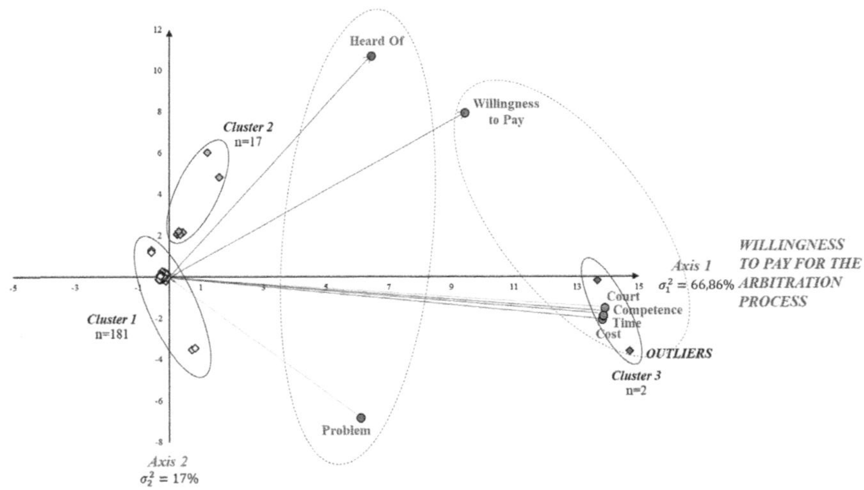

Figure 3. Demand Segmentation.

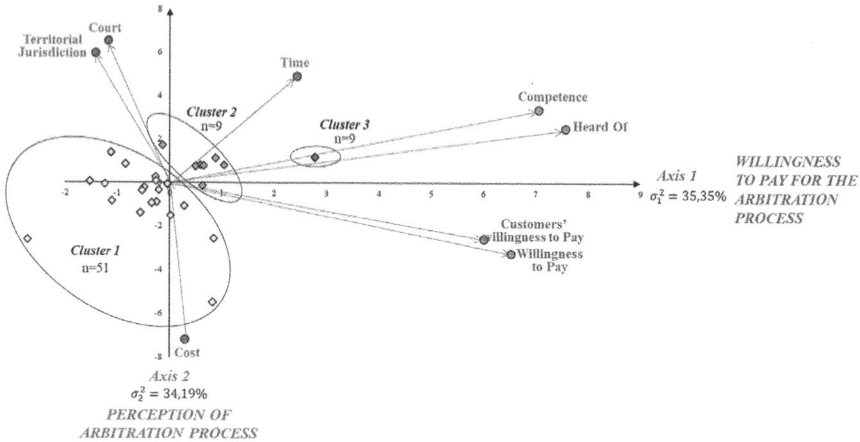

Figure 4. Supply Organizations' Segmentation.

cluster analysis according to Ward's method and is described in the same factorial structure of the HJ-Biplot.

In order to identify a possible competitive advantage, the investigation analyzed the asymmetry between willingness to pay of supply chains and the directly observable difference between the values that demand and supply chains would be willing to pay for resolution via arbitration, all of them measured as a percentage of the total trip cost. Tables 1 and 2 were prepared using the percentages associated with supply chains and demand. Table 1 illustrates how much supply chains and demand would be willing to pay for arbitration. The demand is willing to pay 1% of the total cost of the trip. Table 2 illustrates the perception of the supply about demand's behavior. The 0% indicates that supply thinks demand is not willing to pay anything for arbitration.

The study analyzed the difference between the values that supply chains considered that demand was willing to pay to use arbitration, compared to the values indicated by the demand. This analysis indicated a possibility of exploitation with creation of value for the demand and revenue increase for the supply chains. Bearing in mind that these are percentages over the total cost of a trip, and the probability of having problems is very low, it means a potential huge gain for the tourism industry.

Regarding arbitration, it should be noted that tourism demand had a greater lack of knowledge about its existence and functioning when compared to hotels and travel agencies, as shown in Figure 5. In fact, only 45% of tourists were familiar with arbitration, while 66% of hotels and 67% of travel agencies had heard about arbitration as an alternative method to solve a dispute. Notwithstanding this lack of knowledge, once arbitration was clarified, it was accepted as a valid solution for dispute resolutions.

Table 1. Supply Organizations' and Demand's Willingness to Pay.

		Demand	
		Cluster 1 + Cluster 2 198 Tourists	
Supply	*Cluster 1* 57 Hotels	0.5%	
	Cluster 2 12 Travel Agencies	1%	1%

Table 2. Supply Organizations' Perception on Demand's Willingness to Pay.

		Demand	
		Cluster 1 + Cluster 2 198 Tourists	
Supply	*Cluster 1* 57 Hotels	0.5%	
	Cluster 2 12 Travel Agencies	1%	0%

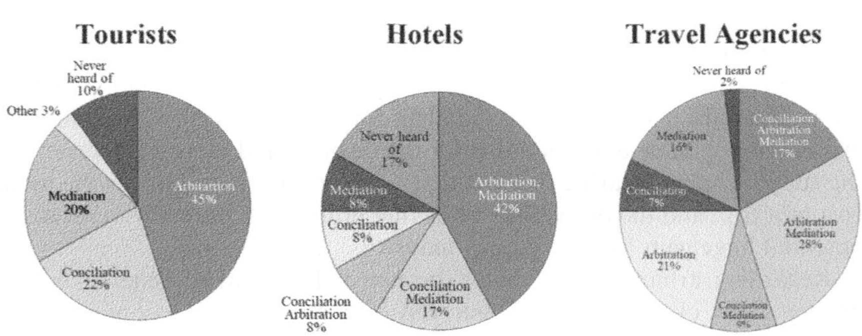

Figure 5. Knowledge About Arbitration.

The most part of the respondents, whether tourists, travel agencies, or hotels, were willing to pay a price in order to ensure a quick and fair dispute resolution through arbitration (Figure 6). This fact reinforces the understanding that the market perceived the advantages of resourcing to alternative instruments for dispute resolutions and valued them.

The positioning of the clusters found with respect to the active variables revealed similar behavior among their individuals. However, the additional variables allowed them to be characterized, with the consequent creation of profiles for the tourism market. It was also possible to detect dissimilarities between hotels and travel agencies (Table 3).

The results show the match between demand and supply chains in the market with respect to fair and speedy resolution of a possible litigation using an arbitration process. Supply chain respondents had an incorrect perception of demand's interests in this area. In this context, it was observed that using the concept of "created value" proposed by Brandenburger and Stuart (1996), considering that the cost of opportunity, as monetary expression associated with the best alternative to use the resource, would have been 1% of the total price of the trip, and that the willingness of tourists to pay for the product with the new configuration was 101%; the associated price would certainly be 101%.

According to Mintzberg (1988), a strategy of differentiation based on support and something that accompanies the product (its support) can serve as a framework for the strategic formulation of companies that want to integrate competitive advantage that comes from submitting their disputes to arbitration. This fact, coupled with the representation that demand makes of a possible situation that requires resourcing to a court to be resolved, in which 97% of the respondents answered that they did not find it possible to have a problem of this type, allowed to state that the results indicated a gap, hitherto unknown, that could be used by supply chains to maximize their results.

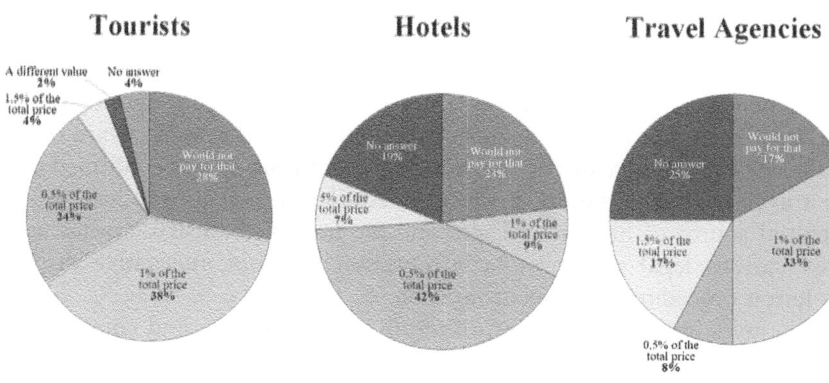

Figure 6. Willingness to Pay for Arbitration.

Table 3. Supply Organizations' Perception on Demand's Willingness to Pay.

	Demand	Supply	
	Clusters 1 and 2 (198 Tourists)	Cluster 1 (57 Hotels)	Cluster 2 (12 Travel Agencies)
Which court?	The court of your region	The court of your region	The court of your region
Which competence?	The Judge is specialized in Tourism	The Judge is specialized in Tourism	The Judge is specialized in Tourism
Which time?	The fastest	The fastest	The fastest
Which costs?	The cheapest as long as it is the fastest	The cheapest as long as it is the fastest	The cheapest as long as it is the fastest
Heard of alternative dispute resolution methods?	Arbitration	Arbitration, Mediation	Arbitration, Mediation
How much will you be willing to pay?	1% of the total cost	0.5% of the total cost	1% of the total cost
In your opinion how much would the client agree to pay?	–	Won't pay for that	Won't pay for that

CONCLUSION

The results obtained indicate that arbitration, as an alternative procedure, is very useful in the general context of dispute resolution since it enables a speedy and specialized decision on existing disputes. It was also observed that, among alternative dispute resolution procedures, arbitration is the most comprehensive and promising procedure to resolve transnational litigations. The results confirm that the agents of the tourism market were aware of the existence of alternative dispute resolution mechanisms. It was confirmed that the travel by individuals that tourism requires to sell its product was a potential source of litigation. Therefore, the tourism industry, insofar as it leads various legal systems to confrontation, requires alternative dispute resolution procedures such as arbitration.

This investigation found that, in terms of performance and competitiveness, tourism organizations benefit from voluntary adherence to international arbitration systems, which is a competitive advantage for the company via differentiation based on support. Indeed, the results unequivocally indicated that tourism demand and supply chains were willing to pay for arbitration processes; however, supply chains wrongly considered that demand was not willing to pay for these processes.

On the other hand, it was observed that although tourism organizations had to market their goods/services at a marginally higher cost because they had resourced to an institutional system of international arbitration, they continued to benefit from higher performance and competitiveness. Indeed, tourism companies were able to raise the price of their products due to the fact that they had resourced to institutionalized systems of international arbitration. This way, they managed to increase profitability, given that the likelihood of a dispute would always be very low. Finally, it was found that the methodological scheme used in this study was adequate for the treatment of the data obtained.

Limitations and Future Research

This study exhibited multiple limitations. The most obvious was the geographic circumscription to Faro District, Portugal. The data collected referred exclusively to tourists who visited the Algarve and tourism companies with headquarters or branches in this region. The second limitation was that only staff of 4- and 5-star hotels were interviewed, excluding hotels classified in other categories, as well as new forms of tourism accommodation, such as hostels and local properties. The last limitation was that the study did not assess other tourism stakeholders, such as car rental companies and passenger transport companies.

Since this study was limited to the Algarve region, a future line of research will be the study of other regions of the country or abroad. Likewise, extending the investigation to other stakeholders can enrich the results and the explanatory capacity of the model now created. Another promising line of research to follow this study will be to analyze the perception of risk by tourists in different markets and how this perception can lead to results that are different from those found. It should be noted that in the Algarve region, the tourists expressed their willingness to pay about 1% of the value of the trip in order to have access to a speedy and fair way of solving a possible dispute. Finally, an assessment of the relationship between international arbitration and the performance of the destination may prove useful for decision-makers, particularly those who govern public tourism policies.

Acknowledgments—This paper is financed by National Funds provided by FCT-Foundation for Science and Technology through project UIDB/04020/2020.

Chapter 14

MARITIME CONNECTIVITY
Resident Population Demand

Alfonso Vargas-Sánchez
University of Huelva, Spain

Juan José Albendín-Moya
University of Huelva, Spain

Fernando Perna
University of Algarve, Portugal

Maria João Custódio
University of Algarve, Portugal

Luís Nobre Pereira
University of Algarve, Portugal

Maria Margarida Santos
University of Algarve, Portugal

Vanessa Oliveira
University of Algarve, Portugal

Abstract: The Spain–Portugal Cross-Border Cooperation Program launched the "Espomar" project in 2017, with the aim of designing a

Managing Destinations: From Theories to Practices
Bridging Tourism Theory and Practice, Volume 14, 193–212

ISSN: 2042-1443/doi:10.1108/S2042-144320240000014014

sustainable and transboundary maritime transport system in the Gulf of Cádiz, which includes the Spanish provinces of Cádiz (its Atlantic coast) and Huelva, together with the Portuguese region of the Algarve. This chapter limits the scope to the Algarve-Huelva connection and the resident population demand. The findings suggest a maritime connection defined and marketed as part of leisure activities. The demand for potential tourism packages created around boat trips appears to be promising. The sun beach, gastronomy wine and culture landscape vectors are the main attractors. **Keywords**: Sea connectivity; resident demand; tourism development; Algarve; Huelva; cross-border cooperation

INTRODUCTION

At the European Union level, Cross-border cooperation (CBC) has had a decisive impact on border region development. The tourism industry has also been a beneficiary, turning borders – a factor that in the past made mobility difficult and separated people – into an opportunity to create popular tourism experiences based on the enjoyment of diversity. This trend has been favored by extensive investment in infrastructure that has dramatically improved accessibility and thus tourist flows.

Portugal's Algarve region and Spain's Huelva province make a good case as this border area changed definitively with the 1991 inauguration of an international bridge over the Guadiana River (their natural border), funded by the European Regional Development Fund (ERDF). However, this cross-border region's connectivity continues to be limited to road transport (cars, trucks, and buses), with no air, rail, or sea links, which has had many implications (Figure 1).

Given this context, the Spain–Portugal CBC Program (a part of the EU's Interreg V Program from 2014 to 2020) approved the Espomar project. This undertaking was launched in 2017, with the goal of designing a sustainable and transboundary maritime transport system for the Gulf of Cádiz. The proposed service includes the majority of the Iberian Peninsula's Atlantic coast and thus Spain's Cádiz province, but the present study dealing with the Algarve-Huelva connection only. The Spanish universities of Cádiz and Huelva, Portugal's University of Algarve and the Public Ports Agency of Andalusia are part of this cooperation network.

Within its designated geographical scope, the Espomar project seeks to improve connectivity, tourist offers, heritage resources' use and natural resource management, and to reduce greenhouse gas emissions. As a result, the border region's natural and cultural potential will be exploited in noninvasive ways, so this project can contribute to improved economic development of all this part of the Iberian Peninsula's coast. To achieve this goal, a series of activities were initiated: a demand study and analysis, design of transport and sustainable vessels, environmental effects investigation, economic analysis, and optimization. This research focused on the project's

Figure 1. Adjoining South Coastal Areas of Algarve and Huelva Provinces.

Source: **Adapted from Google Maps, https://www.google.com/maps/.**

first activity (demand study and analysis) and the resident population in the Algarve region and Huelva province. This chapter is thus a synthesis of reports on these two territories, covering the possible creation of a maritime passenger transport line between them.

This investigation sought to explore the potential demand for an Algarve-Huelva line (under the existing circumstances when it was performed, which could have been affected by the unexpected COVID-19 pandemic and the uncertainty about the new scenario post-coronavirus) as a part of the Espomar initial phase, which would end with an assessment of the transport line's feasibility and economic impacts. The line's potential users are the resident population in the respective areas and "floating" population of tourists. There are significant destinations on both sides of the international border, particularly along the coast where most of the resident population is concentrated. This chapter focuses on the resident community.

To support the whole Espomar project, its quantitative study had two aims. First, to estimate the resident population demand and its temporal distribution (holiday periods vs the rest of the year). Second, to characterize the demand profile through variables such as tourism motivation, time preference, type of travel, single ticket price, journey duration, and preferred arrival and departure ports.

MARITIME CONNECTIVITY DEMAND ANALYSIS

CBC is currently a key strategy used to promote and support more sustainable and competitive regions, improving territorial cohesion by reducing disparities in adjoining regions' levels of social and economic development. European Union has focused on the potential added value of innovative dynamic interactions between endogenous strengths and cross-border opportunities. Various political, social, and cultural development projects have transformed many EU border areas into open landscapes from both a physical and symbolic point of view (Prokkola, 2010). The funding of projects to achieve CBC aims has established more functional economic regions and stimulated more extensive integration within the European Union (Prokkola, 2010).

Support for CBC Programs have also recently focused on "helping to ensure that borders are not barriers, bringing Europeans closer together, helping to solve common problems, facilitating the sharing of ideas and assets, and encouraging strategic work toward common goals" (European Commission, 2019, p. 1). According to Medeiros (2019), the Interreg or European Territorial Cooperation Programs have been the most important EU financial tool for CBC. Particularly in southern Europe, the main objective is to develop quite basic cross-border accessibility, especially where this is nonexistent. The bridge over the Guadiana River is the main connection between the two areas under study (the Algarve and Huelva), in which the first Community Initiative Program INTERREG I (1990–1993) – CBC was used to finance the accesses to the bridge, and the ERDF, with equal responsibilities for both countries, to fund the bridge itself.

Despite these investments, European citizens still perceive various barriers when crossing borders to neighboring countries. According to the report "Overcoming Obstacles in Border Regions" (European Union, 2016), 53% of respondents feel that legal and administrative barriers are the most significant obstacles they face when crossing borders, while 38% mention language barriers and 32% difficult physical access. In addition, public authorities' weak interest in working together and overcoming economic disparities were each reported by 29% of respondents, sociocultural differences by 20%, and a lack of trust by 12% (European Union, 2016). The cited report further reveals that 32% of respondents across a border at least once a week, and their reasons for crossing borders are, in the first place, leisure (almost two in five respondents), followed by the purchases of goods and/or services and visits to friends and family. Employment, business, or educational motivations are mentioned much less often by respondents (European Union, 2016).

According to Medeiros, CBC "is a process that requires the collaboration of one or more entities from at least two different sides of an administrative borderline (normally a country), to achieve a certain (common) goal" (2019, p. 5). Wood and Gray (1991) also suggest that cooperation is when a group of independent stakeholders faced with a problem in a given area is committed to establishing interactive processes and applying shared rules, standards, and structures in order to make and act on decisions related to that area. In addition, these studies characterize collaboration: Stakeholders are voluntary members who share interests or aims; these individuals have a distinct view of different aspects of the same problem or use different approaches; group members apply a constructive approach; they share structures, rules, and standards; their cooperation refers to a problem in a given domain; and the group is searching for solutions capable of inducing change.

The present research adopted a blend of the conceptualizations of Wood and Gray (1991) and Jarillo (1993). In this fashion, cooperation was understood as a joint initiative of independent organizations (public and/or private) that voluntarily share structures, rules, and standards not only to resolve jointly shared problems but also to induce innovation. The latter refers specifically to the development of new products and/or services or to the relevant production processes or another key aspect. The results obtained must also be greater than the sum of individual contributions, and the necessary cooperation is not achieved by mergers but by the creation of a strategic network of independent organizations.

Despite the current interest in and awareness of CBC's positive effects, many collaborations have resulted in failure and contributed to participants' frustration. The reasons for the failure of CBC have been investigated. According to Medina-Muñoz and García-Falcón (2000), major causes of failed cooperation are a lack of trust, participation, commitment, and coordination, as well as deficiencies in the quality of communication and information exchanges and a failure to use conflict resolution techniques. Bramwell and Lane (2000) further identified difficulties in the enormous complexity and slowness of processes involving a wide range of stakeholders. Hall and Jenkins (1995) highlight the impact of groups who refuse to participate for fear of losing influence and power or because they distrust the other partners involved. The challenges identified can be mitigated or emphasized by the socioeconomic, cultural, administrative, and political environments of societies in which this type of inter-organizational relationship is usually operationalized.

Saretzki, Wilken, and Wohler (2002) argue that CBC has better chances of surviving if the organizations involved provide complementary resources, define a support structure, and set up incentives for participation. The cited authors also consider complementary resources important, namely, when the organizations concerned have real or potential internal resources that complement the resources present in other organizations. In other words, resource sharing needs to create positive synergy effects.

The most essential assumptions made in terms of CBC's underlying structure are the avoidance of opportunistic conduct by any organizations involved and the development of mutual trust among participants. In this context, organizations in many cases must share their existing knowledge, which implies potential danger as this exchange of knowledge could be carried out only in one direction. Finally, without incentives, that is, in the absence of win-win solutions, stable forms of cooperation cannot be maintained between organizations. The organizations in question need to realize that their participation will result in increases in effectiveness and efficiency in their different areas of operations in order for these entities to become interested in participating in CBC.

Within this framework, Nilsson, Eskilsson, and Ek (2010) consider tourism and destination development as a relatively new part of regional development strategies, and accordingly also part of cross-border region-building processes. In the European Union, the Interreg Program has been the funding instrument most extensively used to encourage most of the CBC projects, including those based on cooperative tourism development (Ioannides, Nielsen, & Billing, 2006). Nilsson et al. note that "these regionalization processes are supported by the European Union as they are regarded as important for the general integration process" (2010, p. 156).

In fact, the European Union has strongly promoted CBC and made the development of tourism products resulting from cooperation between neighboring regions a priority (Więckowski & Cerić, 2016). In this context, the Spain–Portugal CBC Program's, Espomar project was approved and launched in 2017. Combining tourism theory and practice, the project's goal is to design a sustainable and transboundary maritime transport system for the Gulf of Cádiz. The system will thus include the Atlantic coast of Spain's Cádiz province, with this chapter covering the investigation on the Algarve-Huelva line, which in the area of influence of the project establishes the link between the two territorially contiguous provinces of two different countries.

Espomar is an innovative example of applied research on maritime connectivity and tourism development. The regions involved share a set of ports situated in close geographical vicinity, which could reinforce territorial contiguity far beyond the current shared road transport. The project seeks to examine the ports' capacity for still unexplored potential maritime transport lines. Espomar focuses on exploiting opportunities (but only in passenger shipping) to develop short sea ferry operations, which are defined by Cottam et al. as "maritime transport for the carriage of people and or their cars, coaches, and trucks on short-haul routes around coastlines of the world's continents" (2007, p. 187).

The increased importance of maritime connectivity to regional economic integration is widely recognized worldwide because this connectivity can generate economies of scale. The relevant literature includes extremely positive assessments, among which stands out Trace et al. observation that "improving connectivity, especially through transport links, is an essential condition for economic growth" (2009, p. 2). Cross-border maritime links help the authorities to develop more effective policies and regulatory infrastructure for the goods and services trade (Tamura, 2016). Therefore, Espomar has sought to emphasize the benefits to recreational, leisure, and tourism services in local communities and the related added value for supply chains. In addition, according to Tamura (2016), maritime connectivity must be based on mutual infrastructure development, which Espomar has included as an assumption in analyses of Huelva and the Algarve.

Sea connectivity can also be a strategic way to apply the principles of self-dependence, self-governance, and self-financing and, as a policy, maritime transport supports commitments to regional sustainable development and competitiveness. An example of relevant good practices was found in a Croatian passenger shipping analysis: "expansion of domestic and foreign networks, diversification of markets, development of maritime infrastructure, and the integration of maritime transport with national and regional development strategies are some of the key tasks that are currently being undertaken" (Cottam, Roe, & Challacombe, 2007, p. 198). The Espomar project and Interreg Program have updated and specifically incorporated this multidimensional approach.

Analyzing sea connectivity's potential requires an examination of demand for coastal passenger shipping and its link with tourism transport needs. According to Alexopoulos and Theotokas, the market for these maritime services "can be considered a derived demand" (2000, p. 5). Namely, the need for coastal shipping transport can vary according to the demand for basic transport that produces these services. The seasonality effect implied by this perspective constitutes a key argument for considering local communities' recreational, leisure, and tourism opportunities as part of any maritime connectivity demand analysis.

However, when the focus is on residents or travelers, Alexopoulos and Theotokas argue that "the characteristics of the transport system can ... [all other things being equal], affect the level of demand" (2000, p. 5); this is based on their analyses of daily visits to Psara by tourists who spent vacations in Chios, Greece. This is why characteristics such as itineraries' schedules and other attributes of maritime transport had to be included in the current research.

Despite the expected benefits of maritime connectivity linked to CBC, the literature recommends that researchers should take into consideration a number of actors. Cottam et al. (2007) and Trace, Frielink, and Hew (2009) suggest the possible costs and problems associated with strategies' implementation (how necessary upgrading infrastructure is and whether to support substantial investment in port facilities); the development of complementarities among regional economies (the Algarve and Huelva) could conflict with national priorities and policies (Portugal and Spain); passenger transport companies and passengers tend to take a rigid approach to maritime passenger shipping (the competition from other modes of travel, especially road transport); and the need to separate short-term gains mentioned in policy recommendations from pay-offs is likely to happen only in the long term, including, when necessary, the costs of fundamental shifts in-laws, customs, or practices.

Lopes and Dredge (2017) also reveal a range of local conditions and structural characteristics that create barriers and opportunities for generating different types of values linked to shore excursions that maritime connections could generate, such as the explored in the current research on the maritime connection between the provinces of the Algarve and Huelva. Furthermore, Santos, Radicchi, and Zagnoli (2019) focus on the necessity to deliver bridging between maritime transport and inshore tourism activities (public/private) as a critical factor to guarantee the contribution the sustainability of destinations, linking maritime operators with the local business and infrastructures. Finally, as a base parameter that remains current, Trace et al. suggest that "before setting a definite timeline for implementing policy measures designed to enhance regional maritime connectivity, the interrelatedness of such measure[s] have to be thoroughly understood" (2009, p. 32).

Study Methodology

To achieve the present investigation's aims, initial exploratory research based on secondary sources was conducted to understand the factors that influence consumer behavior in the areas under study. The review of papers published on residents' perceptions and attitudes in both geographical areas also includes its influence on the competitiveness of the destinations and targeting win-win relations between the port and the inland territory's activities (Perna, Custódio, & Oliveira, 2018; Renda, da Costa Mendes, & Oom do Valle, 2011; Silveira, Santos, & Perna, 2018; Vargas-Sánchez, Porras-Bueno, & Plaza-Mejía, 2011, 2014; Vargas-Sánchez & Riquel-Ligero, 2020), together with some reports issued by consultancy companies about the maritime connection between them, were used as sources for variables identification and questionnaire design. Nevertheless, to this respect, it is relevant to bear in mind that the framework established by Espomar project's aims decisively conditioned the content of this research, of a descriptive nature, and focused on residents' opinions.

The methodology followed is detailed in the number of steps: The research at the preliminary information collection stage was limited to obtaining the necessary data on the resident population in the Huelva province and

Algarve region. The National Statistical Institute's databases were used in both countries in order to define clearly the population under study, including its geographical distribution and gender and age profile. As to the goal and the sample, a set of aims was established regarding defining the characteristics and frequency of travelers' movements between Huelva and the Algarve. These objectives included determining residents' predisposition to travel by boat, the importance of various factors or conditions to intentions to use that specific means of transport, the range of travel time between the harbor of departure and of arrival and quantification of the maximum price to be paid. In addition, the data needed to include the duration of trips preferred destinations and informant profiles. Next, the population was defined from which the data would be collected. The study universe or target population was limited to individuals living in areas within municipalities at a distance by road of no more than half an hour from both areas' sea coast. Within these parameters, residents over the age of 18 in Huelva and 25 in the Algarve were considered potential informants. The reason for this age difference is the availability of population data by age ranges.

The optimal sample size was calculated (Table 1), with an acceptable 95% confidence level and a 0.04897 margin of error. A sample of 400 residents was considered the optimal size for Huelva. For the Algarve, the accepted margin of error was 0.0425, so this sample had to be larger: 531. Due to budget and

Table 1. Sample Definition.

Characteristics[a]	Huelva Province	Algarve Region
Total population	519,596	441,929
Population over 18 years old	410,461	329,876
Male population	257,571	211,884
Female population	262,025	230,045
Residents of legal age (>18 years old) in selected municipalities[b]	351,926	312,469
Optimal sample size	400	531
Margin of error	0.04897	0.0425
Level of confidence	95%	95%
Variance	0.25	0.25
Fieldwork[c]	January–February 2018	March–April 2018

[a]Simple random sample.
[b]Those at a drive not exceeding half an hour from the seacoast.
[c]Survey method was in-person interviews on the street.

time constraints, the number of separate surveys to be conducted had to be reduced since initially that the maximum level of error allowed was expected to be lower. Although the findings were based on data collected in person-to-person surveys conducted with the two samples of 400 and 531 residents, an online survey was also carried out in Huelva at the same time as an in-person survey to provide further corroboration of the results from a larger number of observations. The total sample's elements were distributed to achieve proportional sampling by municipalities. The distribution was based on their number of inhabitants and distribution by gender and age.

The above discussion presented the framework on which the questionnaire was designed to achieve the defined aims. This measurement instrument was structured as shown in Table 2.

After the questionnaire was developed to collect the desired data regarding the resident population, the fieldwork was designed, starting with the selection and training of the survey teams and the way in-person surveys were to be administered. This step was followed by contracting a company in the case of Huelva or qualified personnel in the case of the Algarve to complete the fieldwork. The person-to-person survey was carried out, in the first quarter of 2018, in various strategic locations of the selected municipalities to obtain the

Table 2. Questionnaire Structure.[a]

Sections	Variables
1. Trips	Frequency; main motivation; time period; number of people
2. Willingness to travel by boat	Predisposition; frequency; main motivation; time period
3. Conditions (by relative importance)	Distance to departure harbor; weather; price; travel time; distance from harbor of arrival
4. Distance to/from the front door	Time to harbor of departure/time from the harbor of arrival
5. Price	Maximum price of a single ticket
6. Travel time	Maximum time for basic transport service; maximum time for recreational-tourist service
7. Destination	Preferred harbor; attraction of tourist package including boat transfer; more attractive tourist products
8. Respondent profile	Gender; age; municipality of residence; education; employment status

[a]Questions were usually closed and single choice (i.e., dichotomous and polynomial); responses were given on noncomparative scales, both nominal and numerical.

diverse profiles required. The researchers involved in the project closely monitored the survey's progress, and questions that arose were resolved immediately.

In the subsequent phase, an Excel database was created to record the data gathered with the questionnaire from the residents. However, its completion was entrusted to the contracted company or trained personnel mentioned previously. The extraction and subsequent analysis of the online survey data had to be carried out using the same Microsoft Office software due to the difficulty of exporting the online database in question to IBM Statistical Package For The Social Sciences (SPSS) statistical software.

The analysis and interpretation tasks were carried out with the support of statistical techniques considered to be in line with the pursued objectives. An SPSS package was used that allows the importation of data recorded as an Excel database. First, a descriptive analysis for each variable was carried out, and then specific variables were cross-referenced and statistically significant differences verified in order to identify different behavioral patterns. Finally, after the survey data was integrated, a detailed report of the results was compiled. This report was made available to the project's partners and management.

Study Results

The results are presented in the same sequence followed to estimate the resident population demand. The first phase was to calculate the size of the potential market (Table 3). The population statistics and the survey data were used to estimate the percentage of those traveling between the two destinations in question and the percentage of residents willing to do this by boat.

In phase two, the residents' intention to travel was estimated based on the total frequencies of trips on an annual basis obtained from the surveys. In

Table 3. Residents – Potential Market Calculation.

Line	Origin	Reference Population[a]	% Traveling to Destination[b]	% Willing to Travel by Boat to Destination[c]	Potential Market (Number of People)[d]
Huelva-Algarve	Huelva	351,926	66.5	86.8	203,139
	Algarve	312,467	67	37.4	78,298

[a]Resident population of legal age residing in the municipalities (latest data available).
[b, c] surveys.
[d] = [a] multiplied by [b] multiplied by [c].

addition, a correction factor defined previously based on the surveys' results was applied. The respondents' intention to travel was thus calculated by considering the relationship between boat crossings' price and duration since, depending on these conditions, not all individuals who are willing to travel by boat are willing to use this new service (Table 4). The next table (Table 5) provides more detail on the seasonality factor. These results reflect the relationship between the residents' intention to travel during the holiday period and the rest of the year based on the survey data.

To estimate the demand, one final step was needed. In this last phase, intention to travel was subjected to a second correction factor. The literature shows that a precedent exists for using purchase intention as a proxy for actual purchase, which highlights the conceptual and quantitative differences between both variables (intention and behavior). Similarly, intention to travel (by boat in this case) is an antecedent of the actual demand that these maritime transport lines are expected to meet. Thus, to provide a quantified assessment of intention to travel, it must be subjected to a correction factor because not all residents who express an intention to travel will actually do so.

Table 4. Residents – Calculation of Intention to Travel.

Line	Origin	Potential Market (Number of People)[a]	Annual Frequency (Number of Times)[b]	Correction Factor (Price/ Crossing Duration)[c]	Intention to Travel (Number of People)[d]
Huelva-Algarve	Huelva	203,139	2.51	0.6322	322,347
	Algarve	78,298	1.16	0.3083	28,000

[a]Taken from Table 3.
[b,c] Surveys.
[d]= [a] multiplied by [b] multiplied by [c].

Table 5. Residents – Intention to Travel Seasonality.

Line	Origin	Intention to Travel[a]	Holiday Period	%	Rest of Year	%
Huelva-Algarve (number of people)	Huelva	322,347	74,462	23.1	247,885	76.9
	Algarve	28,000	20,048	71.6	7,952	28.4
	Total	350,347	94,510	27.0	255,837	73.0

[a]Data used were the results shown in the last column of Table 4.

This correction factor consists of the intensity of the link (correlation) between purchase intention and actual purchase, which depends on the type of goods or services involved (if it is a rational or impulse purchase). When purchase decisions are based on impulses, a weaker correspondence will exist between purchase intent and behavior. The results of purchase intention analyses are helpful to form a clearer idea about a product's acceptance in the market, especially regarding more negative aspects since low purchase intention indicates that the product offered is unlikely to be successful. However, positive results obtained from this type of analysis indicate that potential consumers are interested in the product even if in the end they do not buy it.

The specific nature of the products under analysis (new passenger shipping lines) meant that no clear reference points were found in the literature. In general, the corrective factors used are usually between 0.5 and 0.8. For instance, one can multiply the percentage of people who answer "sure, I would buy it" by 0.8 and the ones that answer "I would probably buy it" by 0.5. However, this approach was not applicable to the present study because the residents' willingness to travel by boat was assessed using dichotomous terms: "Yes" or "No."

Thus, as already mentioned, the study focused on the more or less rational or impulsive nature of purchases of this product. The upper end of the gap (0.8) corresponded to purely rational purchases, while the lower end of the gap (0.5) was associated with purely impulsive purchases. A fundamentally rational component was, therefore, attributed to purchases of tickets to travel by boat to the destinations considered. Normally, this would reflect pre-determined decisions within residents or tourists' travel plans. Given that the residents' main motivation to travel is leisure and that their trips' economic aspects are relatively insignificant, responses for intention to travel were placed on a scale with reference to a factor of 0.7 (rational with some nuances), which was termed the "intention correction factor."

In this part of the analysis, Hennessey, Yun, and MacDonald (2009) work was considered relevant as it estimated a factor of 0.787 for first-time tourists and 0.8824 for those who repeat their visit to the Canadian island of Prince Edward. This approach is consistent with Fitzsimons and Morwitz's (1996) findings that repeat purchases show a more consistent relationship between intention and behavior. The present study had to take into consideration that the product is a new means of transport (boat) and leisure activities will significantly influence the level of demand, particularly during holiday periods. The methodology needed to reflect closer proximity to the first-time visit scenario to the detriment of the repeat visit scenario.

The intention corrective factor was thus applied to this research on a hypothetical new means of maritime transport between Huelva and Faro in order to consider the association between potential consumers' attitudes and behaviors. This link is influenced over time by an exogenous variable identified as "climate," more specifically defined as the weather that tourists experience when they visit either destination. Researchers widely recognize that climate and/or weather greatly influences behaviors, especially when these involve natural resources or outdoor activities. This impact was identified and quantified by Becken and Wilson (2013) in their study of changes in tourists' routes, schedules, and activities in New Zealand as a result of weather conditions.

In the present research, the variable "wind speed" was used as a proxy for the weather that passengers will experience during crossings. Given the type of trips considered in this study (leisure), a comfort scale was established based on observations of wind speed as measured on the Beaufort scale. The assumption was made that, under specific wind and/or sea conditions, potential users of this means of transport will opt not to take the ferry boat. This assumption was reinforced by the survey results for the areas involved, which confirmed that residents consider the weather one of the most significant adequate conditions before deciding to travel by ship.

The data analysis revealed that this variable has importance of 7.9 and 8.6 for the Algarve and Huelva, respectively, on a scale of 0–10 (minimum to maximum significance). Thus, the assumption, with a certain degree of subjectivity, was refined – to include a scenario in which suitable weather scenario was established as a wind speed equal to or less than four on the Beaufort scale, meaning that the wind blowing at a speed not greater than 16 knots and small waves becoming longer with fairly frequent "white horses."

A second scenario was defined in which these suitable conditions cease to exist due to higher wind speeds and the resulting deterioration of sea conditions, which would lead potential users to cancel plans to travel by boat. More specifically, the nonexistent travel scenario is a wind speed equal to or greater than five on the Beaufort scale, that is, winds blowing at a velocity equal to or greater than 17 knots and, at best, moderate waves taking a more pronounced long form with many white horses and the chance of some spray.

The statistics on the Windguru website (www.windguru.cz) were consulted for Faro and Espigón de Huelva. The observed frequencies of wind speed are between 0 and 4 and between five and 12 for the three locations of reference listed in Table 6.

The above data suggested the need for a second correction factor after purchase decisions are made, based on the guaranteed comfort based on the percentage of acceptable windy days (winds up to speed four on the Beaufort scale). The level of comfort was differentiated according to summer months (the holiday period), and the remaining months (the rest of the year). Thus, the average values for this comfort factor between July and September (holiday period) would be 0.973 and between October and June (the rest of the year) 0.855.

These results supported the conclusion that, in order to determine potential demand, estimates needed to be corrected based on the two factors:

Table 6. Wind Speeds (Beaufort Scale) for Huelva and the Algarve.

Location	Wind Speed (Beaufort Scale) 0–4	5–12
Huelva (Espigón)		
July–September (holidays)	98.9%	1.1%
October–June (rest of year)	87.5%	12.5%
Faro		
July–September (holidays)	95.7%	4.3%
October–June (rest of year)	83.5%	16.5%
Huelva-Faro mean		
July–September (holidays)	97.3%	2.7%
October–June (rest of year)	85.5%	14.5%

Note: Data collected between January 9, 2006 and October 8, 2018.

corrected demand during the holiday period (estimated intention to travel multiplied by an intention correction factor of 0.7 multiplied by a comfort factor of 0.973 equals intention to travel multiplied by 0.6811); and the corrected demand for the rest of the year (estimated intention to travel multiplied by an intention correction factor of 0.7 multiplied by a comfort factor of 0.855 equals intention to travel multiplied by 0.5985).

This study further found that, in addition to comfort (climate), the actual materialization of the corrected demand will depend on four other conditions: distance from the origin to port, price, duration of crossing, and distance from destination to port. Since these four variables are currently unknown, no additional restrictions on demand were considered beyond those already taken into account. Once the ports (departure and arrival), prices, and boat ride duration are known, new correction factors may need to be applied. The respective demand estimates corrected with these factors would be as shown in Table 7.

Table 7. Estimated Demand (Number of People).

Line	Period	Intention[a]	Demand
Huelva-Algarve	Holiday[b]	94,510	64,371
	Rest of year[c]	255,837	153,118
	Annual total	350,347	217,489

[a]Taken from Table 5.

[b]Correction factor of 0.6811.

[c]Correction factor of 0.5985.

Thus, given a 100-day vacation period, the Huelva-Algarve line could serve a daily average of 643 passengers and/or residents. For the rest of the year (260 days), that average would decrease to 588. Synthesis of other qualitative variables contemplated in the present study is listed in Table 8, with a comparison of the results obtained for Huelva residents with those for Algarve residents.

CONCLUSION

Following the entry of Portugal and Spain into what is now the European Union, in 1986, CBC has increased exponentially between these countries, in general, and between the Algarve region and the Huelva province of the Autonomous Community of Andalusia, in particular. This trend started almost from zero, and it has mostly been supported by European territorial cohesion policies and Interreg initiatives funded by the ERDF. Increased

Table 8. Profiles of Future Algarve/Huelva Maritime Line.

Variable	Huelva	Algarve
Motivation (a)	Leisure	Leisure
Time preference (single)	In morning	In morning
Type of travel	Families and/or groups	Small groups (2–3 people)
Single ticket price (basic transport service)	No more than €30 (similar to buses) (b)	No more than €30 (willingness to pay prices slightly higher than for buses) (b)
Duration of crossing (basic transport service)	2 hours maximum	3 hours maximum
Duration of crossing (recreational-tourist service)	3 hours maximum	4 hours maximum
Preferred departure port	Huelva (Muelle de Levante)	Portimão, followed by Faro
Preferred arrival port	Faro (c)	Huelva
Appeal of potential tour packages created around boat trip	Very high	High

Notes: (a) The sun beach, gastronomy wine and culture-landscape highlight as main motivations. (b) €20–22. (c) no clear concentration in responses: designated alternative ports located for residents on windward side.

CBC has favored the progressive integration of neighboring regions that, for centuries, had lived with their backs turned to each other.

This process has not been simple or fast, but CBC has generated opportunities for socioeconomic development through the progressive elimination of all kinds of barriers, especially physical-geographical ones, mostly due to investment in new transport and communications infrastructure. Among these structural links are the three international bridges built in just over two decades between the Huelva province and the Portuguese regions of the Algarve and the lower Alentejo. The first one was constructed across the Guadiana river in 1991 and the other two over one of its tributaries, the Chanza river, in 2009 and 2012. These bridges have transformed the formerly almost impenetrable border areas into permeable spaces that favor commercial, tourism, and cultural exchanges, among other links.

Connecting infrastructure such as this has been successfully combined with the creation and support of programs reinforcing cooperation, such as Euroregions (Alentejo-Algarve-Andalusia) and Eurocities (Ayamonte-Castro Marim-Vila Real de Santo António). This CBC has fostered dynamic initiatives focusing on economic activities and innovations that support, among other areas, productive sectors. In this context, the only existing way connections can be made is by land, so a maritime link that satisfies the criteria of environmental sustainability could be another element that stimulates exchanges between Spanish and Portuguese territories, especially tourism. However, this project's economic viability needs to be studied.

Viability analysis of a maritime passenger transport line between the Algarve and Huelva necessarily included research on potential demand among the resident population.

This evaluation had to be both quantitative, including the seasonal distribution of demand, and qualitative, characterizing potential clients' profiles. To obtain the required data to make estimations, a series of calculations were made following a sequence of three phases. One, based on quantifications of the reference population, the existing potential market (the number of residents) was estimated by considering both the percentage of this population traveling between the two destinations and the percentage willing to do this by boat. Two, based on this potential market, intention to travel was calculated by applying first the annual frequency (number of times) with which individuals travel between the two destinations and a correction factor related to the crossing's price and duration. Three, to estimate the final demand, a correction factor was applied to the results obtained in the previous phase for intention to travel in order to get a closer estimate of residents' actual purchase behavior (traveling by boat, in this case), because prior research has shown that not everyone who expresses an intention to travel will actually do so. Table 9 summarizes all the calculations and their results, which allowed conclusions to be drawn about the different consumer behaviors expected in both border areas.

Table 9. Process Followed in Estimating Demand (Number of People).

Phases	Huelva	Algarve	Total
1. Reference population	351,926	312,467	664,393
1.1 Percentage traveling to destination	×0.665	×0.67	–
1.2 Percentage willing to travel by boat	×0.868	×0.374	–
2. Potential market	= 203,139	= 78,298	281,437
2.1 Annual frequency	×2.51	×1.16	–
2.2 Correction factor (price and duration)	×0.6322	×0.3083	–
3. Travel intention	= 322.347	= 28.000	350.347
3.1 Correction factor (intention and comfort)	×0.6176*	×0.6576*	–
4. Total estimated demand (annual)	= 199,075[a]	= 18,414[b]	217,489
4.1.1. Total estimated demand (holiday period)	50,716[c]	13,655[d]	64,371
4.1.2. Estimated daily demand (holiday period, 100 days)	507	136	643
4.2.1. Total estimated demand (rest of year)	148,359[e]	4,759[f]	153,118
4.2.2. Estimated daily demand (rest of year, 260 days)	570	18	588

Notes: *Average for entire year calculated ex-post (see Table 5).
[a]74.462 × 0.6811 + 247.885 × 0.5985.
[b]20.048 × 0.6811 + 7.952 × 0.5985.
[c]74.462 × 0.6811.
[d]20.048 × 0.6811.
[e]247.885 × 0.5985.
[f]7.952 × 0.5985.

Differences in residents' behaviors in the two areas in question are made apparent by the following major four findings. First, while the percentage of the respective populations traveling on the other side of the border is practically identical, Algarve residents show a significantly lower predisposition to travel by boat (less than half that reported by Huelva residents). Second, the frequency with which Algarve residents travel to Huelva province is slightly less than half of what Huelva residents report regarding trips to the neighboring Portuguese region. Third, in the Algarve region, residents are much more sensitive to the crossings' price and duration than Huelva residents are, but, regarding the second correction factor linked to intention to travel and comfort depending on wind and sea conditions, the difference is negligible. Fourth, while Huelva residents' demand would have a stable distribution throughout the year (23.1% during the holiday period and 76.9% the rest of the year), Algarve residents' demand is significantly concentrated in the holiday period (71.6%, with only 28.4% the rest of the year).

As the study results show, the distribution of demand between both periods is almost the opposite in the two areas. Huelva's residents estimated daily demand is even slightly higher outside the summer vacation period. Nonetheless, this demand must be adjusted to reflect the extent that weather conditions other than wind speed and the configuration of the transport services offered may fall short of consumers' expectations and thus could make the boat trip less attractive.

As a result, the Huelva residents' contribution to total demand is quite unbalanced, with the boat service having a much greater appeal among residents on the Spanish versus the Portuguese side. The explanation offered for this marked difference was not supported with empirical research, as this issue was not included in the present study's objectives. However, based on an informed understanding of both areas, the main reason could be that, in more than 90% of the cases, the residents' trips were motivated by leisure needs, and the Algarve has attractions of which most are known to appeal to Huelva residents throughout the year. In contrast, the tourism and/or recreational products offered by the Huelva province are not as well known among Algarve residents, which may limit the frequency and timing of their trips to this part of Spain. The Algarve's residents show an extremely clear preference for traveling to Huelva for summer holidays, which undoubtedly will be an important factor to consider when the possibility of commercially exploiting this line is being considered.

The greater aggregated demand during the holiday period in combination with the predominant leisure motivation and appeal of potential tourism packages suggests that the summer months as the most suitable time of year for any future implementation of the planned maritime line. Those who organize the plan's execution must also consider the competition from land transport (Table 8) and carefully select the right departure and arrival ports. The latter locations must consider tourism flows since the viability of this maritime connection will depend on its ability to attract tourists as an essential, additional source of demand.

The proposed passenger shipping line between the Algarve region and the Huelva province is part of a strategic effort to improve cohesion between the two cross-border regions. This study's results support the conclusion that this project could successfully avoid most of the limitations identified in the literature review. Nevertheless, the two major restrictions on the transport line's feasibility remain the large investment needed to start the ferry service, which was not considered in the present analysis, and the greater and immediate direct private financial benefits expected from other means of transport, particularly road vehicles. Conversely, this project requires neither changes in laws, customs, or practices nor large investments in upgrading infrastructure or actual port facilities to allow passengers to board and disembark from ferries in Algarve and Huelva ports. In addition, regional, national, and European stakeholders already have a comprehensive understanding of policies and measures designed to enhance regional maritime connectivity.

Finally, due to the approach of this research, mainly focused on the gathering of primary data on the demand side in two specific areas to provide support to the whole Espomar project and back future public policies and decisions, the applied research prevails over the basic search of theoretical implications on the consumer (resident) behavior related to a particular maritime connection. The chapter lacks this latter kind of orientation, which can be considered a limitation, although, at the same time, it is an opportunity for further discussions and future publications aimed in this direction. Nowadays, this is even more necessary due to the possible change in perspectives and decision criteria of the consumers when facing travel decisions.

Acknowledgment—This research is part of the Espomar project, which is cofinanced by the ERDF through the Interreg Program V-A Spain–Portugal (Spain–Portugal CBC Program) 2014–2020. This study is also financed by National Funds provided by Foundation for Science and Technology through project UIDB/04020/2020.

RESEARCH AND KNOWLEDGE MANAGEMENT

Chapter 15

COCREATING INNOVATION
A Study on NEST Living Lab

Ana Cláudia Campos
University of Algarve, Portugal

Abstract: Innovation is critical to the tourism industry. Living Labs (LLs) are innovation ecosystems which are becoming popular in tourism. LLs are a model of open innovation characterized by a networked approach to innovation. Local communities and businesses are usually excluded from processes of decision-making concerning the management of tourism. LLs enable local communities' participation in the sustainable development of tourism, thereby providing them the opportunity to increase their share of benefits. This chapter focuses on a case study of an LL recently created in Portugal and discusses LLs as an environment for tourism innovation where the local community is stimulated to explore competencies to cocreate value. **Keywords**: Living lab; innovation; cocreation; community competencies; NEST; tourism ecosystem

INTRODUCTION

Innovation is a critical driver for economic and social development, as societies evolve toward knowledge-intensive arrangements (Compagnucci, Spigarelli, Coelho, & Duarte, 2021), and concurrently sustainability issues reach global awareness (Hossain, Leminen, & Westerlund, 2019). Over the last decade, the innovation paradigm has embraced an open innovation model based on the creative capital of individuals, simultaneously aware of the need for sustainability and environmental protection.

The open innovation model is the result of competitive marketing dynamics which require increased market sensitivity (Chesbrough, 2006; Martinez, 2014; OECD/Eurostat, 2005). Consumer creativity and digital

Managing Destinations: From Theories to Practices
Bridging Tourism Theory and Practice, Volume 14, 215–233

ISSN: 2042-1443/doi:10.1108/S2042-144320240000014015

communication platforms are of paramount importance. Consumers must be active participants in innovation processes, acting as external sources of knowledge and expertise. Collaboration in research and development activities is due to a combination of their capacity to create new ideas, use digital technologies, and willingness to dialogue with firms.

Living Labs (LLs) are networks that foster stakeholder inclusion, participation, and collaboration in processes for innovation able to change social and economic dynamics and are considered important drivers of participative development (Compagnucci et al., 2021). Residents are now recognized as key stakeholders in the democratic processes that make critical decisions about sustainable development of tourism (Arabska, Shopova, & Dimitrova, 2014; Kantsperger, Thees, & Eckert, 2019; Thees, Pechlaner, Olbrich, & Schuhbert, 2020).

Recent research in tourism has shown the potential of LL to bring changes to territorial dynamics by harnessing regional and local competencies, creativity, and resources, as well as forge synergistic networks for the benefit of service providers, tourists, the territory, and the wider community (Pucihar, Malešič, Lenart, & Borštnar, 2014; Šker & Floričić, 2019; Thees et al., 2020). There is evidence, however, that in some cases, residents' competencies to collaborate in innovation are still low and that education is needed (Priday & Pedell, 2017).

This chapter provides a study of NEST, an innovation networked ecosystem recently created in Portugal that enhances sustainable development of tourism by providing community inclusion and enhancing competencies to innovate. The NEST CEO was interviewed to provide an in-depth perspective on how the involvement of the community and local entrepreneurs in innovation is increased. Insights from the case analyzed will provide the basis for other destinations to enhance community participation in a globally competitive and digitally connected world.

OPEN INNOVATION AND COCREATION

Firm innovation is increasingly interacting with the wider business environment either by accessing and absorbing ideas from the outside or outsourcing innovations to the market. In this open process, innovation is the outcome of the relationships that the firm develops with competitor companies, start-up firms, suppliers, research organizations, customers, public authorities, funding agencies, technology consultants and enablers, media, and conference organizers. This distributed process creates specific challenges and demands from people (Podmetina, Soderquist, Petraite, & Teplov, 2018).

As increased competitive market forces emphasize the importance of knowledge, both industry-based and customer-based, open innovation is now popular (Bénézech, 2012). Technologies develop more rapidly and customers constantly evolve in their knowledge, tastes, and preferences, thus product or process obsolescence becomes critical and a source of business disadvantage in the market. Thus, responsiveness, adaptability, and learning are driving forces of open innovation.

A world characterized by borderless markets and digital technologies has enabled new channels to communicate innovation, more competent and creative consumers have become creators of knowledge, and companies increasingly engage in collaborative practices (Martinez, 2014). Currently, firms recognize that creativity lies everywhere, inside and outside the firm, and that communication with the outside world is what drives business innovation, performance, and advantage by facilitating this process of knowledge creation, flow, and absorption. Open innovation means permeable boundaries and emergent ecosystems for innovation. This evolution in innovation thinking and practice has led to more complex network environments by integration of a variety of participant actors, most often facilitated by informal, fluid, and flexible interactions (Bénézech, 2012).

Cocreation challenges the premise of the centrality of firms in the creation of value. Instead, a co-creation mindset stipulates that value is cocreated by actor interaction (Prahalad & Ramaswamy, 2004; Vargo & Lusch, 2004). An open innovation model depends on intensive interaction between parties so that knowledge can be collaboratively created, shared, and absorbed. Willingness to cocreate and joint the value creation process is increasing among consumers, and their involvement in innovation is moving to the product design stage (Martinez, 2014). Cocreation views consumers as empowered and active creators, rather than passive receptors of value.

LL: Open Innovation and Cocreation Environments

LLs are variedly described as a methodology, a system, a concept, an instrument, an environment or ecosystem, as shown in Table 1. *The European Network of Living Labs* (ENoLL) considers an LL as a user-centered open innovation ecosystem based on a systematic user cocreation approach, integrating research and innovation processes in real-life communities and settings. LLs have the ability to connect users, technology, and businesses into an open innovative development process in real-life environments. A recent definition by Hossain et al. (2019) includes the virtual environment.

The ENoLL platform indicates that there is interest in developing LL in contexts such as business, education, climate and the environment, agriculture, creative industries, smart cities and design. LLs may be physical or virtual and take many forms depending on area of intervention, user participation (designer/tester/validator), or the geographical basis (region/network innovative structure) (https://enoll.org/about-us/).

Table 1. Definitions of Living Lab.

Author(s)	Definitions
Cigir (2019, p. 569)	"a user centred open-innovation research concept and environment that integrates concurrent research and innovation processes"
Leminen et al. (2012, p. 7)	"physical regions or virtual realities in which stakeholders form public-private-people partnerships of firms, public agencies, universities, institutes, and users all collaborating for creation, prototyping, validating, and testing of new technologies, services, products, and systems in real-life contexts."
Mabrouki et al. (2010, p. 66)	"a research methodology for innovation that challenges the whole research and innovation process in real-life conditions by human, social, cultural, organizational and institutional aspects, and has an impact on sustain- able service, business and technology development"
Thees et al. (2020, pp. 7, 13)	"a user-centric or also user-driven innovation environment, where managers, creators, and all kinds of users interact in developing innovations with a social and economic impact" "cooperative environments that follow a user-driven and cross-stakeholder approach in order to provide a shared space for experiments and tests of innovative products and services in a real-life environment".
Westerlund and Leminen (2011, p. 23)	"a real-life test and experimentation environment where users and producers cocreate innovations".

LL as Cocreative Innovation Environments

LLs apply the open innovation model (Arabska et al., 2014; Boes, Buhalis, & Inversini, 2016; Haukipuro et al., 2018; Hossain et al., 2019; Schuurman, De Marez, & Ballon, 2016; Šker & Floričić, 2019; Veeckman, Schuurman, Leminen, & Westerlund, 2013) and disseminate tools and methodologies across stakeholders (Jernsand, 2019), adopting a cocreation approach to business (Guimont, Lapointe, & Sévigny, 2017; Leminem, 2013), which brings users to the center and involve them throughout the innovation process (Lupp et al., 2021). A cocreation logic is central to LLs, as they assume innovation is a collaborative process rather than a single-inventor result (Schuurman, De Marez, & Ballon, 2013) and is thus accelerated by inclusion of external sources in the early stages (Haukipuro et al., 2018; Westerlund & Leminem, 2011). LLs are defined as an *approach* and *environment* to develop innovation. As an approach, they are characterized by bottom-up or community-led processes of innovation as studied and tested in real-life contexts with real users (Åström, Ruoppila, Ertiö, Karlsson, & Thiel, 2015). The approach involves activities which require application and integration of actors' resources, thus actors, activities, and resources are key elements of LLs (Leminem, 2013).

LLs are environments created to foster innovation. They are called "ecosystems" or "constellations" (Guimont & Lapointe, 2016) that facilitate and synchronize interaction between actors who participate in the design, testing, validation, development, and commercialization of new technologies, goods, services, and experiences in a real-life environment (Cigir, 2019; Compagnucci et al., 2021; Guimont & Lapointe, 2016; Šker & Floričić, 2019). They are also arenas for development of innovative business models and systems (Jernsand, 2019) that by integrating users help reducing risk of failure (Arabska et al., 2014). LLs are usually composed of physical facilities to gather people in co-ideation and co-designing processes, but also provide the context for test beds, tools for networking, knowledge management platforms, and human resources (García-Guzmán, Fernández del Carpio, De Amescua, & Velasco, 2013). Currently, due to technological developments and the pervasive role and influence of digital technologies, LLs are increasingly making use of digital platforms to drive innovation and actors' inclusiveness in the innovation processes from start to end (Cigir, 2019; Guimont et al., 2017; Leminen, Westerlund, & Nyström, 2012). Accordingly, definitions of LLs have integrated the digital sphere as one of the dimensions of these dynamic systems (Hossain et al., 2019).

Community's Participation in LLs as Open Innovation Spaces and Critical Competencies

Community participation in processes of decision-making in areas that impact public as well as private life is now perceived determinant for democratic societies to thrive (Åström Ruoppila, Ertiö, Karlsson, & Thiel, 2015), and consequently open models for participation are growing in pace (Pasher, Herzog, Harir, Turjeman-Levi, & Zhiqiang, 2018). Domains such as political participation to sustainable development, rural or urban development, computing and technological enhancement of social spaces, education, and tourism are leading to the dissemination of the new citizen-driven paradigm

(Åström et al., 2015; Baran & Berkowicz, 2020; Boes et al., 2016; Leino & Puumala, 2021; Mabrouki, Amirat, Chibani, Fernandez, & De La Cruz, 2010; Pasher et al., 2018).

Communities are host to diverse human capital and resources, which places them at the heart of innovative change. International awareness of this leads public officials to make efforts in residents' inclusion in the innovation process, stimulating destinations to develop innovation ecosystems (Compagnucci et al., 2021). The management of innovation processes is now a mix of bottom-up and top-down strategies to deliver innovation outcomes to society, given that need of awareness and new ideas are captured by a bottom-up approach, while policy, innovation validation and implementation requires top-down direction (Baccarne Schuurman, Mechant, & De Marez, 2015; Leminem, 2013, 2015; Lievens, Schaffers, Turkama, Ståhlbröst, & Ballon, 2011).

Digital technologies and new media are changing social life and boosting community participation and engagement. Despite acknowledgment that participation is to some extent dependent on the nature of innovation projects (Lupp et al., 2021), public pressure for community participation is increasing (Baccarne et al., 2015). LLs are ecosystems that depend highly on operant resources and competencies to participate, cocreate, and innovate (Åström et al., 2015; Leino & Puumala, 2021). However, these are lacking among both public government organizations and community members (Baccarne et al., 2015; Kantsperger et al., 2019). Leino and Puumala (2021) hold that residents' participation in cocreative processes differ according to their capacity and resources and thus that inclusion is sustainable as long as accessibility to participation is ensured. Thees et al. (2020) emphasize that residents are diverse because they have different roles, interests, and motivations which may influence their degree of participation in collaborative processes.

In the context of tourism and community involvement in tourism development, a community's composition and different attitudes of residents not directly interested in or affected by tourism may influence their motivation to participate in a collaborative process (Kantsperger et al., 2019). Destinations in underdeveloped countries most affected by tourism development and international investment face problems of enclave tourism and economic leakages (Li, 2017). Regardless of a community's interest in the development of tourism, their participation is compromised by ineffective regulations and complex stakeholders' relationships, but also by residents' passivity and lack of competencies. LLs enable stakeholders to break away from long accepted hierarchies that hinder coordinated participation and acquisition of competencies to develop and consequently to participate (Thees et al., 2020). Many challenges emerge in this specific context and individual competencies to address them are found decisive (Podmetina et al., 2018).

Table 2. Personal Competencies for Development and Innovation.

Sources and Context	Information, Knowledge, and Learning	Motivation, (Passion, Effort, Engagement, Involvement, Interest)	Creativity and Critical Thinking	Technological Capacity	Meaningful Goals	Entrepreneurial Attitude/ Initiative
Haukiputo et al. (2018) Education, aviation, health care	X	X		X		
Jernsand (2019) Higher education and sustainable tourism	X	X	X		X	X
Kantsperger et al. (2019) Tourism development	X	X				X
Leino and Puumala (2021) Urban development	X	X		X	X	
Li (2017) Tourism development and community participation	X	X	X	X		X
Lupp et al. (2021) Nature management and planning	X	X			X	
				X		

(Continued)

Table 2. *(Continued)*

Sources and Context	Information, Knowledge, and Learning	Motivation, (Passion, Effort, Engagement, Involvement, Interest)	Creativity and Critical Thinking	Technological Capacity	Meaningful Goals	Entrepreneurial Attitude/ Initiative
Mabrouki et al. (2010) Rural communities						
Pasher et al. (2018) Smart cities and CoP	X	X	X		X	X
Pucihar et al. (2019) Autonomous mobility	X	X		X		
Pucihar et al. (2014) Tourism services	X			X		
Thees et al. (2020) Tourism destination management and governance	X	X		X	X	X

Note: Conference of the parties (CoP).

Table 3. Interpersonal Competencies in Development and Innovation.

Sources and Context	Interpersonal Competencies						
	Shared Vision and Power (Decision-Making)	Altruism	Communication (e.g., Feedback)	Cooperation/ Collaboration	Knowledge Sharing	Trust	Relationship and Social Interaction
Haukiputo et al. (2018) Education, aviation, health care			X	X			X
Jernsand (2019) Higher education and sustainable tourism	X		X			X	X
Kantsperger et al. (2019) Tourism development			X	X	X		
Leino and Puumala (2021) Urban development	X		X			X	X
Lupp et al. (2021) Nature management	X			X		X	
Pasher et al. (2018) Smart cities and CoP	X	X	X	X	X	X	X
Pucihar et al. (2019) Autonomous mobility				X		X	
Pucihar et al. (2014) Tourism services			X	X			
Thees et al. (2020) Tourism destination management/governance	X		X	X			X

Note: Conference of the parties (CoP).

Resident competencies important in leading or participating in successful collaborative initiatives and participation are provided in Tables 2 and 3. Competency means an individual's attributes that are necessary for effective performance of a role important to successful open innovation, such as knowledge or intellectual abilities (Grama & Royer, 2013). Competencies have been described as a mix of skills and abilities important to innovation, the former more dependent on the individual's training, experience and knowledge transfer, while the latter are closely related to innate attributes (Podmetina et al., 2018). Conceptualization of competencies as skills and abilities allows discerning the *what* and the *how* of an individual performance.

LLs require competencies to participate in innovation at two levels. Individuals need "personal" competencies while groups require "interpersonal" competencies (e.g., trust). Mabrouki et al. (2010) found that residents' poor technical knowledge compromised their capacity to reflect on the needs of a service-oriented architecture to support a rural innovation ecosystem. Thus, residents' personal competencies impact on ability to cocreate innovation with the other stakeholders. In collaboration, competencies that facilitate relationships between stakeholders become key to success.

Overall, interpersonal competencies are relevant in collaborative environments and contexts (Podmetina et al., 2018), where common goals and objectives are set for the stakeholders. So, competencies that facilitate relationships among them are involved and required. A number of studies (Table 2) have highlighted that interpersonal competencies, such as communication, knowledge sharing, interaction, and trust enhance the opportunities to contribute to innovation.

LLs Applications in Tourism

Implementation and research on LLs in tourism is growing (Arasbka et al., 2014). The fragmentation of the tourism industry at destination level (Guimont et al., 2017) demands new ways of stimulating innovation. LLs are seen as arrangements that enhance destination management organizations' (DMO) innovation capabilities through facilitation of collaborative action among stakeholders. LLs increase the opportunities of successful innovation by helping to overcome recognized weaknesses of the tourism industry, such as dimension, diversity of businesses, and seasonality (Guimont et al., 2017). LL arrangements are diverse and develop according to specific purposes and expected life spans (Leminem, 2015). Participants perform different roles. Leminem (2015) categorizes LLs according to participant type (e.g., as utilizer-driven or enabler-driven), leading to different purpose, action, outcome, lifespan, and network development. In tourism, participants include public officials such as the regional and/or municipal authorities and policymakers, investors, start-up entrepreneurs, universities, community members and associations, and tourists. Tourism has a territorial base which encourages the creation of LLs in the pursuit of sustainable development (Pucihar et al., 2014).

LLs are appropriate to support the competitiveness of local tourist companies and the local community, or even an entire tourism area (Šker & Floričić, 2019). This is especially important when residents are acknowledged underrepresented in processes involving the development of tourism (Kantsperger et al., 2019; Thees et al., 2020). Effective implementation of LL requires specification of the area's customer segments, the delimited territory, and even the type of tourism desired. Thus, context and information and communication technology (ICT) infrastructure, networks, leadership, governance, and cocreation activities have been considered important components of tourism LL (Boes et al., 2016; Hossain et al., 2019). LL implementation combines a bottom-up with a top-down approach to innovation (Boes et al., 2016), allowing the community and tourists to contribute to creative outcomes, that translate into the proposition "working with tourists" instead of "working for tourists" (Guimont et al., 2017).

LLs facilitate and drive tourism development, competitiveness, and sustainability and generate societal benefits, as well as benefits for the tourism businesses (Cigir, 2019; Guimont & Lapointe, 2016; Thees et al., 2020). Societal benefits include consideration of economic, social, and environmental values, and increase of general awareness to problems associated with negative impacts brought by touristic activity (Boes et al., 2016; Hossain et al., 2019). Moreover, LLs impact the community in a positive way by including residents' social capital in the innovation process. Residents' and micro entrepreneurs' inclusion in LL process is increasingly dependent on digital technologies (Stankov & Gretzel, 2020) and the dissemination of smart infrastructures which allow the sharing of information and of the innovative solutions (Boes et al., 2016). For businesses, LLs are an ecosystem that facilitate access to market opportunities and innovative tourism experience concepts. For the tourists, LLs are environments that stimulate interaction with tourism providers, empowerment, and participation in cocreative activities, contributing to greater capacity to respond to ever changing tourists' needs (Guimont et al., 2017; Šker & Floričić, 2019; Westerlund & Leminem, 2011).

Study Methods

This case study explored NEST as an approach to tourism innovation and the community's competencies to cocreate value for the development of tourism. As a qualitative and explorative research, data triangulation was adopted as a way to collect information from various sources. Primary data were obtained from an in-depth interview conducted with the NEST CEO, as the management's perspective on community's competencies was considered critical to explore and uncover opportunities for competence development in this particular setting. Secondary data were collected from NEST's website, available documents provided by the CEO, and a case report on NEST (OECD, 2020).

The interview took place on June 17, 2021, for approximately 2 and half hours using the Zoom platform. The interview focused on the following issues concerning NEST: (i) origins and characterization, (ii) objectives and approach to innovation, (iii) perspective on community's competencies and participation in innovation. Data were subsequently content analyzed according to the predefined research topics.

Analysis

Origins and characterization: NEST was created in May 2019 with the founding members Turismo de Portugal, ANA Airports, Brisa Via Verde, Google, Microsoft, Millennium BCP, NOS and BPI, a group of private and public organizations. These entities formed a legal private nonprofit association, with organizational bodies such as General Assembly, Fiscal Council and Board. According to the CEO (*RA*), NEST was first conceived by the Portuguese Board for Tourism, *Turismo de Portugal*, as a network innovative structure to support the strategic development of tourism in Portugal. NEST was defined by *RA* as:

> ...an innovation center to boost the development of innovation and dissemination of digititalization in the tourism value chain, (...) a knowledge facilitator to SMEs that links companies, start-ups, and learning institutions in order to enhance the potential to create a competitive advantage to the sector.

Despite NEST's national mission and scope, headquarters are located in Covilhã, inland central region of Portugal. NEST uses its premises and physical spaces to organize processes of collaboration and to allow people to gather in creative discussion. NEST has developed virtual platforms and tools to connect the tourism stakeholders nationwide, allowing access to innovations to all regions of Portugal. At the macro level, NEST focuses on establishing clustering effects through virtual knowledge exchange in a group of technology-based firms, providing digital tools, taking on the moderating role in the coordination of the tourism ecosystem for increasing the pace of innovation; at a micro level, it seeks to establish trust among stakeholders involved in innovation projects.

NEST's approach to innovation and objectives: As an innovation ecosystem, NEST's mission is to promote innovation and technology in tourism through development of new business ideas and project experimentation that may benefit the whole tourism system. Moreover, it fosters the transition of companies to the digital era using three areas of intervention:

innovation culture, digital knowledge, and innovation projects. A culture for innovation is proactive, ahead of changing trends, and uses digital tools intensively. NEST facilitates knowledge creation and transfer through technological tools which connect people and allow application of skills. In the words of *RA*, in Portugal

> . . .to innovate in tourism involves working with those already operating in the market but that need to adapt their business models and strategies to the environment, and also with the new entrants and start-ups, somehow supporting the pursuit of economic development based on the growth of an economy emerging from these small companies (. . .) but above all those start-ups focused on digital developments. These companies have had the largest growth worldwide and are able to bring contributions to the economy. (. . .) The digital world is allowing the democratization of many good things that leverage growth in many areas and especially entrepreneurship.

The structure and development of tourism in Portugal requires public intervention and direction, and many areas of tourism need improvement. An innovation culture itself needs enhancement, and innovation programs based on knowledge transfer, gaps and needs of the sector are designed to invite stakeholders' participation. NEST innovation is focused on SMEs who contribute approximately 70% of active businesses in tourism. As *RA* claimed:

> These types of companies are special, as they are poor in terms of innovation culture, investment capacity and association focus. Moreover, they are dispersed.
>
> For these small companies, most of the times, innovation means access to solutions, and not the creation of the solution itself. If only they could use 5% or 10% of the current technological capacity, and through it be more agile, improve communication, and provide employees with the means to be more autonomous. . .
>
> Organizations need to train digital skills and implement processes because their business models are being challenged by small companies and entrepreneurs, and many innovations come from the outside and other sectors. Currently the world is 'phygital' in the sense that the virtual dimension is as much as real as the physical one. For a company to exist, it needs to have a digital expression. Those small companies that aren't digitally connected, e.g., as to Google maps, just don't exist to the tourist. Many of them don't have their own website.

NEST faces four types of challenges: social, economic, environmental, and technological. Social and environmental priorities combine the tourists' needs of seamless experiences at the destination with the community's expectations toward good use of regional resources and positive environmental impacts from tourism; economic ones focus on business responsiveness to changes in the market environment and decisions; and technological priorities are based on systems and tools for information crossing and continuous improvement. In addition to the core value proposition, NEST makes available a selection of resources to facilitate the process of inclusion of key actors in the industry so that knowledge flows through the network and providers, community, and tourists jointly participate in the shaping of a competitive and sustainable tourism (Table 4).

NEST's perspective on community's competencies and participation: NEST is an initiative that meets the European Union's objectives of boosting European countries development of artificial intelligence, computer science, and cybersecurity in order to reduce dependence on the United States and China. The tourism industry is dependent on foreign technological companies, weakening European countries' opportunities to accelerate and lead innovation. Thus, facilitating access to technological tools that stimulate people's capacity to innovate is among its priorities.

> Digitalization is worthless if not responding to people's needs. It helps valuing time and work, by allowing computers to do what they were designed for. It then frees people to be sensitive, insightful, and creative. In the paradigm, assumptions seem to be neutral, specifically in what concerns entrepreneurship democratization; wide access to digital tools leads to people empowerment and autonomy, however their usefulness depends on people capacity to consult and interpret trends, analyze information, and build communication and content. These technological skills have major impacts on organizations. (...) Employees are able to contribute with insights and new knowledge, they can do more things and add value to processes. Digitalization is an operational model and those not able to work with it are excluded from development and competition; thus, digitalization involves an entrepreneurial attitude.

The link between digitalization and entrepreneurship is apparent, but innovation is facilitated by the nature of tourism itself.

> Being an entrepreneur in tourism means that the person is curious and in search of a lifestyle, and this has been stimulating the creation of new tourism offerings. The experience of tourism is immersive, informal and aspirational, and that may be the reason why tourism is so intertwined with social life. People's well-being can be found in tourism, because of the job opportunities that creates, or the protection of traditions and cultural assets, such as gastronomy or local crafts.

Table 4. NEST's Resources and Activities.

NEST Resource	Description	Objectives
Think Tanks	Inline discussion sessions with experts on Gastronomy and Restaurants, Hospitality, Tourism Activities, and Points of Interest.	To study the Portuguese tourism market and identify main challenges, needs, and opportunities within the tourism innovation processes.
Tourism Toolkit for small and medium enterprise (SME)	Apps list to optimize firms' digital strategy	To facilitate firms' access to event platforms and self-assessment tools
Future Lab Hotel	Partnership project with Setúbal Higher School for Tourism. Open space to start-ups or similar organizations aiming at product and service development and experimentation	To become a national reference in innovation in tourism; To develop Minimum Viable Products (MVPs); To disseminate knowledge through the network; To integrate students and teaching staff in advanced digital and innovation experiences
Upgrade Program	Online questionnaire and training program for skills improvement	To assess the initial level of quality of tourism professionals To provide training in skill improvement for tourism
Top 20 Startups	Listing of most innovative start-ups in tourism	To create awareness of innovations in tourism To disseminate knowledge through the network
Keep Innovation in Tourism (K.I.T) Webinars	Keep Innovation in Tourism is a program of 30 free webinars to occur during the year	To empower professionals in innovation in tourism businesses

(Continued)

Table 4. *(Continued)*

NEST Resource	Description	Objectives
Webinar Tourism Trade Talks	Talks and discussion in webinar format, debating the 4 main outbound markets for Portugal	To discuss changes and activities in each market, aiming at relaunching tourism growth To discuss key challenges ahead
Apps and Solutions for COVID-19	App list to support management in times of reopening of tourism activity, covering topics such as business efficiency and events	To create awareness among the network to the need of responding to challenges in times of crisis
Startups	Shared information on events, resources and measures available to prepare for the future of tourism	To support start-up companies in the context of global crisis
Events	National/international • UNWTO Global Rural Tourism Startup Competition • Mind the Tourism • Tourism Creative Factory	To promote solutions and development in the most vulnerable areas To stimulate creative solutions for sustainable innovation in tourism

Source: NEST's website; https://issuu.com/nestportugal/docs/report_think_tank_points_of_interest; *R.A.* interview.

Innovation requires the skill to detect opportunities in the market environment, the creative capacity to imagine solutions and implement them, therefore, enhancing an entrepreneurial attitude is seen as vital to successful innovation. Many skills are not the result of formal learning. However, technical knowledge is recognized important to understand the relationship, e.g., between a business idea and a market segment. Inland territories are considered less dynamic areas in terms of communities' willingness to mobilize for innovation; thus the will, the strategy, and the interest to be part of a process of change need to be developed and stimulated through residents' joining a collaborative network for the competitive development of tourism. For NEST, the understanding of the entrepreneurial potential supports the inclusion of local producers (e.g., the local cheese producer) in the process of innovation in tourism. There are, however, barriers to accelerated innovation:

> Tourism is not yet developed in inland regions; development is still in its early stages, the ecosystem for innovation is taking its first steps, so it doesn't benefit from past learning experiences. It takes a little longer.

Overcoming barriers to innovation is a challenge addressed by implementing collaborative processes for auscultation of which areas need to be developed in order to create value for tourism. Community's entrepreneurs are critical in this dimension (Figure 1).

> NEST has been creating collaborative networks, for instance using Slack software, connecting approximately 1500 people of the sector for exchanging ideas and information sharing, allowing people to closely dialogue, among them companies, start-ups, people from

Figure 1. Community's Competencies.

> education institutions, and technology developers, linking people which are dispersed geographically. T^+ is such an example of a digital tool that has been communicated and collectively discussed. NEST works with several living labs to create an information circuit fed by the needs and feedback from the real world. (. . .) NEST supports tourist-focused projects that enhance the destination experience by helping detecting gaps and bringing from society's internal strengths the solution providers. NEST connects the system, but only collaboration among all allows closing the gaps in the ecosystem.

CONCLUSION

The importance of innovation in tourism is paramount, and competitiveness is today strongly linked to the principles of social and environmental sustainability (OECD/Eurostat, 2005). The forces that have shaped more informed societies also influenced the way that innovation is created. Competition is global, but also without boundaries, and businesses must connect to cutting-edge industries, companies, and creative individuals that have the skills to meet people's needs. Innovation is open to all willing to cooperate, and cocreation of value is becoming the norm (Prahalad & Ramaswamy, 2004).

LLs are a way to cocreate innovation, as they develop as environments connecting ecosystems' stakeholders in real life contexts. Analysis of current examples shows LL are both a concept for innovation creation and a way to develop and implement it that is based on an ecosystem view where all stakeholders gather and commit to apply and exchange resources for the benefit of all. LLs accommodate different formats and arrangements, however networked structures contribute to disseminate the logic of open innovation for the sustainable development and competitiveness of tourism through supporting acceleration of innovation, enhancement of digital capacity, and the development of an entrepreneurial culture.

There are barriers to innovation that need to be addressed, and one way to help overcoming them is by developing an ecosystem for innovation that links stakeholders and creates synergies among them. NEST is an example of a network that connects tourism regions and the entire national tourism system for the competitiveness of the sector. This network evolved as an ecosystem which transcends its original territorial basis, despite performing as a strong facilitator of regional development of tourism. This combination of a macro and a micro approach to innovation in tourism is striking in the case of NEST, which seems to benefit from *Turismo de Portugal*'s vision of a sustainable and competitive tourism sector in the country. The tourism sector is, in Portugal as in many other countries, composed of small companies which depend on entrepreneurial will and interest in the creation of new tourism offerings. An ecosystem for innovation is pivotal in the case of tourism due to the contributions of the various areas that combine to facilitate memorable destination experiences. In this regard, NEST's role is believed to generate positive impacts to the region where physical headquarters are located but also to the whole tourism sector, which is disseminated in the national territory. NEST is best defined as a networked structure for innovation in tourism which links the stakeholders at macro and the micro levels conducting experiments in real-life contexts, and forms an ecosystem composed of physical and virtual dimensions. This LL ecosystem is helping small companies to connect to technological firms and start-ups for synergetic creativity and design, to the benefit of regional and national tourism, and tourists visiting the country in search of seamless and appealing experiences.

Other destinations facing similar challenges, such as boosting tourism sustainability and competitiveness at the local, regional, and national levels, may likewise gain from implementation of this twofold approach to tourism development, as positive results to stakeholders' participation and performance of the tourism sector are now becoming apparent. The community and entrepreneurs that together shape the destination experience are called to participate in the innovation process of tourism. NEST ecosystem goes hand-in-hand with digital transformation and the entrepreneurial people that cocreate value for tourism, thus enhancement of competencies to propose creative solutions and scale up innovation is a unifying goal of NEST's management and stakeholders.

More research is needed concerning LL and innovation ecosystems in the context of tourism development, sustainability, and competitiveness. Good examples from practice are also needed to shed light on positive dynamics among stakeholders and results. The aggregation of macro and micro perspectives adopted by NEST is original and worthy of attention, as most cases adopt one to the exclusion of the other. How to further improve stakeholders' competencies for innovation and how to refine innovation ecosystems and LLs to better address community's requirements to successfully cocreate innovative solutions still need research and discussion.

Acknowledgment—This chapter is financed by National Funds provided by FCT–Foundation for Science and Technology through project UIDB/04020/2020.

Chapter 16

CONCLUSION

Noel Scott
Edith Cowan University, Australia

Manuela Guerreiro
University of Algarve, Portugal

Patrícia Pinto
University of Algarve, Portugal

This book draws its inspiration from papers submitted to the t-Forum 2020 conference. The t-Forum is about transfer of knowledge/intelligence to and within tourism. The papers selected for this volume contribute to the transfer of knowledge about destination management. Thus, it is appropriate to draw some conclusions about destination management research and also about knowledge transfer. Two issues have been identified for further discussion of the many that arise from the chapters in this volume. The first concerns developing skills and expertise in destination management and the second bridging tourism theory and practice.

Reflections on Destination Management

The chapters in this volume were chosen to reflect aspects of destination management. They are divided into five sections: strategy, new markets, stakeholders, infrastructure and research, and knowledge management. Within these sections, cross-cutting themes include the importance of social and environmental sustainability, diversity of cultural perspectives and treatments, and integration of governance.

Managing Destinations: From Theories to Practices
Bridging Tourism Theory and Practice, Volume 14, 235–238

ISSN: 2042-1443/doi:10.1108/S2042-144320240000014016

At the end of this discourse, it appears that destination management is broad in scope and beyond the expertise of any one person. Development of a destination involves knowledge of markets, stakeholders, cultural and even religious views, infrastructure, environmental sustainability, and governance. If there is any glue that holds research on destination management together, collaboration, cooperation, and competition issues stand. Several chapters in this book discuss the perspectives of stakeholders such as residents, business owners, and government organizations. Surprisingly, explicit study of collaboration and cooperation in the tourism literature is limited. The existing research tends to identify problems with little discussion of good practice.

This may be because destination management requires soft skills and expertise, not easily reduced to specific instructions. This gap in the literature may be an interesting topic for further conferences on bridging tourism theory and practice.

Bridging Tourism Theory and Practice

As noted earlier, many of the papers in this volume were presented at The Tourism Intelligence Forum (t-Forum) 2020 Global Conference held in Algarve, Portugal, from 4 to 7 November 2020. The t-Forum and its conferences are all about the transfer of knowledge/intelligence to and within tourism. Accordingly, reflecting upon the chapters of this book, attempts are to be made to discuss research that crosses boundaries and transfers knowledge.

Internal or External Perspectives

The chapters in this book may be classified as adopting an *internal* or *external* perspective on tourism. The former covers senior tourists' booking behavior (Chapter 7), profiling tourists in terms of well-being (Chapter 6), residents' attitudes (Chapter 8), and their support towards tourism (Chapter 10), World Heritage Site and local residents' views (Chapter 11), or developing innovative tourism products (Chapter 16), and cross-border tourism (Chapter 4).

From an *external* perspective, some chapters are about the interface of tourism and other sectors, including marine transport (Chapter 15), ports (Chapter 14), urban and regional planning in Italian rural areas (Chapter 2), northern Italian cities (Chapter 3), Venice (Chapter 5), law and arbitration (Chapter 13), and solar power (Chapter 12).

In the tourism literature, the internal tourism perspective is more common. The focus is fully on tourists, stakeholders, or developing new and innovative products. The common focus of tourism researchers on this stand is understandable as attention is drawn to familiar phenomena and research topics.

On the other hand, an external perspective can be rewarding in that it places tourism in context where its relativity or standing among a wider system can be revealed. For example, tourism is important in developing new forms of transportation, planning for urban spaces, developing new forms of electric power, or examining the effect of the law on tourism services. The external perspective gives one more of a sense of insight, whereby one can see tourism as just one part of a larger picture. This perspective can be invigorating.

Tourism researchers often focus narrowly on tourism, its impacts, customers, stakeholders, planning, and so on. Adopting a wider systemic viewpoint, whereby tourism is just one part of a transportation system, electricity supply grid, or city planning outcome appears more sympathetic to the concept of sustainability.

Transfer of Perspective Knowledge

Internal focused knowledge is communicated to a tourism audience. This is the type of intelligence that is much discussed in the knowledge transfer literature. External focused knowledge transfers wisdom to other sectors and to people who are not academics (tourism business owners) or to people who are not involved in tourism at all. In comparison to the internally focused knowledge examining only tourism, external perspective knowledge begins by being generated in an integrated way.

Several chapters of this volume view tourism as only a part of the phenomena being studied. A port (Chapter 14) on new maritime shipping route is not exclusively tourism infrastructure and must be studied from a wider perspective. The law of arbitration of tourism contracts is part of a wider body of legal knowledge. The discussion of urban and regional planning in three of this book's chapters highlights how tourism is intertwined with other issues of competitiveness, power, and politics. The use of solar power in tourism in Croatia is part of a wider narrative that goes far beyond its borders. These chapters integrate, not separate, tourism and are, therefore, arguably more successful in bridging theory and practice. This is because they integrate intelligence of tourism along with other knowledge in addressing a practical situation.

The chapters with an internal view face more of a problem in transferal of knowledge. Such chapters must find an audience that is interested in its findings. This is a more difficult task than beginning with that audience being involved in their research. The chapters with an internal view are more specialized, with perhaps a narrower audience.

Implications for the Future

In 2021, 10,752 tourism journal articles were published. It is not known what percentage of these articles are read by the industry. Is knowledge generated for recycling internally? What would distinguish an internally focused journal article that is read by industry from that which is ignored? One likely explanatory factor that would predict knowledge transfer is industry's involvement or at least interest in the internally focused research topic. The internally focused chapters in this volume are all industry-based research.

This analysis suggests that bridging tourism theory and practice requires more than translation or interaction between different groups. An external perspective would suggest that it is the research problems to be studied that also require agreement. In addition, it may be useful to take a wider perspective than only tourism in studying some problems. In planning for future conferences and books, it is necessary to work with industry to identify broad problems of which tourism is just a part. It may also increase transferability of knowledge by asking industry participants what type of research they are interested in learning about.

The topic of destination management, as unfolded in this volume, involves experts from many areas. Therefore, a conference on bridging tourism theory and practice, focused on destination management, would need to involve urban planners, transportation, and legal experts. One of the criticisms of tourism conferences is that they do not often attract academics or industry participants from outside tourism. It is easy to explain this as "their" lack of understanding or interest. An alternative explanation is "our" lack of suitable research or knowledge. "We" need to focus on bridging theory and practice, not just in better communication of findings, but by a better understanding of research needs. Interestingly, tourism researchers need to develop their soft skills in engaging industry participants. For this purpose, hospitality and tourism education should exert efforts to connect theory and practice. Is tourism education, especially postgraduate programs, excessively theory-based with limited connections with reality? Should tourism education providers require that students immerse in the industry while completing their degrees? What else should be done so that students can complete their academic degrees with an in-depth understanding of the tourism industry? These topics deserve attention from future tourism conferences on bridging tourism theory and practice.

REFERENCES

Abu-Al-Fida, I. (1850). *Countries calendar.* Dar Sader.

Acevedo, C., & Nohara, J. (2004, September 10–11). Consumer behavior in tourism: A study with senior tourists [Paper presentation]. In *Construções Teóricas no Campo do Turismo. Anais do II Seminário de Pesquisa em Turismo do Mercosul,* Brasil. Retrieved from https://www.ucs.br/site/midia/arquivos/17-consumer-behavior.pdf

Adams, J., Khan, H. T. A., Raeside, R., & White, D. (2007). *Research methods for graduate business and social science students.* SAGE Publications Ltd.

Agapito, D., Valle, P., & Mendes, J. (2013). The cognitive-affective-conative model of destination image: A confirmatory analysis. *Journal of Travel & Tourism Marketing, 30*(5), 471–481.

Agarwal, S. (2002). Restructuring seaside tourism. The resort lifecycle. *Annals of Tourism Research, 29*(1), 25–55.

Agenzia per la Coesione Territoriale. (2013). *per la Coesione Territoriale, Agenzia. "Strategia nazionale per le Aree interne: Definizione, obiettivi, strumenti e governance (Documento tecnico collegato alla bozza di Accordo di Partenariato trasmessa alla CE il 9 dicembre 2013)" [National strategy for internal areas: Definition, objectives, tools and governance (Technical document linked to the draft Partnership Agreement sent to the European Commission on 9 December 2013)].* Rome.

Ahmed, Z. U. (1991). The influence of the components of a state's tourist image on product positioning strategy. *Tourism Management, 12*(4), 331–340.

Akın, M. (2015). A novel approach to model selection in tourism demand modeling. *Tourism Management, 48,* 64–72.

Akis, S., Peristianis, N., & Waner, J. (1996). Residents' attitude to tourism development: The case of Cyprus. *Tourism Management, 17*(7), 418–494.

Al-Baladi, A. G. (2010). *Dictionary of Hijaz landmarks.* Dar Makkah for Publishing and Distribution – Al Rayyan Corporation.

Alén, E., Nicolau, J. L., Losada, N., & Domínguez, T. (2014). Determinant factors of senior tourists' length of stay. *Annals of Tourism Research, 49,* 19–32. doi:10.1016/j.annals.2014.08.002

Alexandrov, V. A. (1984). *Russia on the far eastern borders: The second half of the XVII century.* Far East Publishing House.

Alexopoulos, A., & Theotokas, I. (2000). Quality services in the coastal passenger shipping sector and its contribution to the development of tourism in small islands. The case of Psara island. In *Proceedings of International Conference on: Tourism in Island Areas and Special Destinations* (pp. 1–11). University of the Aegean.

AlGassim, A. A., & Abdulaziz, A. A. A. (2021). The effect of COVID-19 on intraregional tourism of gulf cooperation council countries. *International Journal of Tourism and Hospitality, 1*(1), 1–13.

AlGassim, A., & Abuelhassan, A. (2021). Effect of COVID-19 on potential tourist's consumption behavior: Evidence from GCC countries. *Journal of Association of Arab Universities for Tourism and Hospitality, 19*(3), 129–144.

AlGassim, A., Barry, S., & McPhail, R. (2012). Exploring management and employees' perspective of HRM practices in Saudi Arabia MNC hotels. *GSTF Journal of Law and Social Sciences, 2*(1), 148–154.

Almeida, A. (2016). Modelling tourism demand in Madeira since 1946: And historical overview based on a time series approach. *Journal of Spatial and Organizational Dynamics, 4*(2), 145–156.

Alpa, G. (2013). L'arbitrato e la Risoluzione Stragiudiziale Delle Controversie. In A. Guido & V. Vincenzo (Eds.), *Arbitrato. Profili di Diritto Sostanziale e di Diritto Processuale* (pp. 17–35). UTET.

Alrwajfah, M. M., Almeida-García, F., & Cortés-Macías, R. (2019). Residents' perceptions and satisfaction toward tourism development: A case study of Petra Region, Jordan. *Sustainability, 11*(7), 1907–1926.

Amendola, G. (2003). *La città postmoderna*. Laterza.

Amione, M. (2018). *Rigenerazione e trasformazione urbana: Torino si reinventa*. Retrieved from https://www.presstletter.com/2018/04/rigenerazione-e-trasformazione-urbana-torino-si-reinventa-di-marta-amione/

Amoamo, M. (2018). More thoughts on core-periphery and tourism: Brexit and the UK Overseas Territories. *Tourism Recreation Research*, *43*(1), 1–16.

Andereck, K. L., Valentine, K. M., Vogt, C. A., & Knopf, R. C. (2007). A cross-cultural analysis of tourism and quality of life perceptions. *Journal of Sustainable Tourism*, *15*(5), 483–502.

Andrades, L., & Dimanche, F. (2017). Destination competitiveness and tourism development in Russia: Issues and challenges. *Tourism Management*, *62*(C), 360–376.

Andriotis, K., & Vaughan, R. D. (2003). Urban residents' attitude toward tourism development: The case of Crete. *Journal of Travel Research*, *42*(2), 172–185.

Anselmi, G., & Vicari, S. (2020). Milan makes it to the big leagues: A financialized growth machine at work. *European Urban and Regional Studies*, *27*(2), 106–124.

Ansoff, H. I. (1965). *Corporate strategy: An analytic approach to business policy for growth and expansion*. New York, NY: McGraw-Hill.

Ansoff, H. I. (1991). Critique of Henry Mintzberg's 'The design school: Reconstructing the basic premises of strategic management'. *Strategic Management Journal*, *12*(6), 449–461.

Ap, J. (1992). Residents' perception on tourism impacts. *Annals of Tourism Research*, *19*(4), 665–690.

Arabska, E., Shopova, I., & Dimitrova, V. (2014). Living labs in integrated agriculture and tourism activities: Driving innovations for sustainable rural development. *The Małopolska School of Economics in Tarnów Research Papers Collection*, *25*(2), 27–35.

Arcese, G., Di Pietro, L., & Mugion, R. G. (2015). Social life cycle assessment application: Stakeholder implication in the cultural heritage sector. In *Social life cycle assessment* (pp. 115–146). Springer Singapore.

Arslanova, L., Agapito, D., & Pinto, P. (2017). Tourism destination image of Russia in the perspective of the Portuguese market. *Journal of Spatial and Organizational Dynamics*, *5*(2), 146–158.

Artemenko, S. V. (2010). *Geographical study of cross-border regionalism. The theory of socio-economic geography: Current state and development prospects*. Rostov-on-Don: SFU Publishing House.

Arvati, P. (2005). Sociale. In B. Sessarego (Ed.), *La Genova del 2005. Quaderni del circolo Rosselli* (Vol. 3, pp. 17–31). La Nuova Italia.

Ashworth, G. J., & Tunbridge, J. E. (2000). *The tourist-historic city – Retrospect and prospect of managing the heritage city*. Pergamon.

Assolombarda. (2016). *Milano Città STEAM. La visione di Assolombarda per Milano*. Retrieved from https://www.assolombarda.it/centro-studi/milano-citta-steam-la-visione-di-assolombarda-per-milano

Åström, J., Ruoppila, S., Ertiö, T., Karlsson, M., & Thiel, S.-K. (2015). Potentials and challenges of a living lab approach in research on mobile participation. In *Proceedings of the 2015 ACM International Joint Conference on Pervasive and Ubiquitous Computing and Proceedings of the 2015 ACM International Symposium on Wearable Computers – UbiComp'15*.

Avkiran, N. K. (2018). Rise of the partial least square structural equation modeling: An application in banking. In N. K. Avkiran & C. M. Ringle (Eds.), *Partial least square structural equation modeling recent advances in banking and finance*. St. Lucia/Hamburg: Springer.

Azores Tourism Observatory. (2020). *Satisfação do Turista que Visita os Açores 2010–2018*. Retrieved from https://datastudio.google.com/reporting/1dVrLaxgEz6D3xDYgJLZp3wZzR6LCBX2X/page/KthH?s=n6XD_YbbNKY

Baccarne, B., Schuurman, D., Mechant, P., & De Marez, L. (2014, June 8–11). The role of urban living labs in a smart city [Paper presentation]. In *The XXV ISPIM Conference on Innovation for Sustainable Economy and Society*, Dublin, Ireland.

Baggio, R. (2008). Symptoms of complexity in a tourism system. *Tourism Analysis*, *13*(1), 1–20.

Bagnasco, A. (1977). *Tre Italie. La problematica territoriale dello sviluppo italiano*. Il Mulino.

Bagnasco, A. (1986). *Torino. Un profilo sociologico*. Einaudi.

Bagnasco, A. (Ed.). (1990). *La città dopo Ford. Il caso di Torino*. Einaudi.

Bakr, S. (1981). *Geographical features of pilgrims' paths*. Dar Tehamah.

Baloglu, S., & McCleary, K. W. (1999). A model of destination image formation. *Annals of Tourism Research*, *26*(4), 868–897.

Balsoud, S. (2009). *The Port of Yanbu and Rabigh between 1840–1954*. Umm AlQura University.

Bankwatch Network. (2019). *Bankwatch Network*. Retrieved from https://bankwatch.org/beyond-coal/the-energy-sector-in-croatia

Baran, G., & Berkowicz, A. (2020). Sustainability living labs as a methodological approach to research on the cultural drivers of sustainable development. *Sustainability*, *12*(4835), 1–16.

Barca, F. (2009). *Agenda for a Reformed Cohesion Policy*. European Communities.

Barca, F., Casavola, P., & Lucatelli, S. (2014). *Strategia nazionale per le aree interne: definizione, obiettivi, strumenti e governance*. Ministero dello Sviluppo Economico, Dipartimento per lo Sviluppo e la Coesione Economica. Unità di Valutazione degli Investimenti Pubblici.

Barney, J. B. (1986). Organizational culture: Can it be a source of sustained competitive advantage? *Academy of Management Review*, *11*(3), 656–665.

Barney, J. B., & Hesterly, W. S. (2007). *Administração Estratégica e Vantagem Competitiva*. Pearson Prentice Hall.

Barrocas, M. P. (2010). *Manual de Arbitragem*. Almedina.

Bassin, M. (1999). *Imperial visions nationalist imagination and geographical expansion in the Russian Far East, 1840–1865*. Cambridge University Press.

Bassio, C., & Matera, F. (2011). Development of renewable energy in sicily. *Ufficio Speciale per il Coordinamento delle Iniziative Energetiche*, *2*, 1–7.

Becken, S. (2019). Decarbonising tourism: Mission impossible? *Tourism Recreation Research*, *44*(4), 419–433.

Becken, S., & Wilson, J. (2013). The impacts of weather on tourist travel. *Tourism Geographies*, *15*(4), 620–639.

Beerli, A., & Martin, J. D. (2004). Tourists' characteristics and the perceived image of tourist destinations, A quantitative analysis – A case study of Lazarote, Spain. *Tourism Management*, *25*(5), 623–636.

Belligni, S., & Ravazzi, S. (2013). Policy change without metamorphosis. The 1993–2011 urban regime in Turin. *Metropoles*, *12*. doi:10.4000/metropoles.4642

Benedettelli, M. (1997). L'arbitrato commerciale internazionale tra autonomia privata e coordinamento di sistemi giuridici: riflessioni in margine al nuovo regolamento di arbitrato della Camera di commercio internazionale. *Rivista di diritto internazionale privato e processuale*, 899–918.

Bénézech, D. (2012). The open innovation model: Some issues regarding its internal consistency. *Journal of Innovation Economics*, *10*(2), 145–165.

Beraldo, L. D. F. (2014). *Curso de arbitragem: nos termos da Lei n. 9.307/96*. Atlas.

Bernini, C., & Cracolici, M. F. (2015). Demographic change, tourism expenditure and life cycle behaviour. *Tourism Management*, *47*(C), 191–205.

Berta, G. (2008). *Nord: dal triangolo industriale alla megalopoli padana, 1950–2000*. Mondadori.

Bertelli, C. (1991). Genova, una città che sembra sospesa. In L. Bellicini (Ed.), *La Città Europea Degli Anni '80* (pp. 743–776). Credito Fondiario.

Bethapudi, A. (2013). The role of ICT in tourism industry. *Journal of Applied Economics and Business*, *1*(4), 67–79.

BITS – Bureau International du Tourisme Social. (1996). *Montreal Declaration*. Retrieved from https://www.aitr.org/wp-content/uploads/2014/04/Montreal-declaration.pdf

Blyablina, A. (2015). *The contribution of guides in developing tourist experiences during historical theatrical tours: The case of Stockholm ghost walk*. Master's thesis, Mid-Sweden University.

Boavida-Portugal, I., Rocha, J., & Ferreira, C. C. (2016). Exploring the impacts of future tourism development on land use/cover changes. *Applied Geography*, *77*, 82–91.

Bobbio, R. (2005, October 17–20). Re-shaping spaces for a new economy. The case of Genoa [Paper presentation]. In *41st ISoCaRP Congress 2005*, Bilbao, Spain. Retrieved from http://www.isocarp.net/Data/case_studies/579.pdf

Boes, K., Buhalis, D., & Inversini, A. (2016). Smart tourism destinations: Ecosystems for tourism destination competitiveness. *International Journal of Tourism Cities*, *2*(2), 108–124.

Böhmelt, T. (2010). The impact of trade on international mediation. *Journal of Conflict Research, 54*(4), 566–592.

Boley, B. B., McGehee, N. G., Perdue, R. R., & Long, P. (2014). Empowerment and resident attitudes toward tourism: Strengthening the theoretical foundation through a Weberian lens. *Annals of Tourism Research, 49*, 33–50.

Bollo, A., & Carnelli, L. (2008). *MiTo Settembre Musica. Indagine sul Pubblico di MiTo*. Fondazione Fitzcarraldo. Retrieved from https://cupdf.com/document/il-pubblico-di-mito.html

Bonfantini, B. (2019). Milano verticale/orizzontale: note sulla qualità urbana nel ciclo trasformativo recente. *Saggi e punti di vista/Essays and Viewpoint. Techne, 17*, 86–91.

Boonpienpon, N., Maneenetr, T., Siriwong, P., & Kovathanakul, D. (2015). Indigenous Islamic food: An ideal product innovation for creativity in cultural tourism: A case study Khao Tung Pla (Thai crispy rice crackers with anchovies). *Mediterranean Journal of Social Sciences, 6*(5), 445–453.

Boorstin, D. (1992). *The image: A guide to pseudo-events in America*. Vintage.

Borrelli, N. (2009). *Governance e Territorialità. Una Griglia per L'analisi*. La città del sole: Napoli.

Borrelli, N., & Bernardi, M. (2019). Sharing economy and sustainability in tourism. In S. K. Gupta, V. P. Bhatt, & A. Vaishnava (Eds.), *A handbook of research on international travel agency and tour operation management. Online travel trade in India: Challenges and opportunities* (pp. 116–132).

Bramwell, B., & Lane, B. (2000). *Tourism collaboration and partnerships – Politics, practice and sustainability*. Channel View Publications.

Bramwell, B., & Sharman, A. (1999). Collaboration in local tourism policymaking. *Annals of Tourism Research, 26*(2), 392–415.

Brandenburger, A., & Stuart, H. W. (1996). Value-based business strategy. *Journal of Economics & Management Strategy, 5*(1), 5–24.

Bria, F., & Morozov, E. (2018). *Ripensare la Smart City*. Codice Edizioni.

Brown, S. (2008). Travelling with a purpose: Understanding the motives and benefits of volunteer vacationers. *Current Issues in Tourism, 8*(6), 479–496.

Brunt, P., & Courtney, P. (1999). Host perceptions of sociocultural impacts. *Annals of Tourism Research, 26*(3), 493–515.

Bryman, A. (2015). *Social research methods* (5th ed.). Oxford University Press.

Buckley, R. (2004). The effects of world heritage listing on tourism to Australian national parks. *Journal of Sustainable Tourism, 12*(1), 70–84.

Buckley, R. C. (2018). Tourism and natural world heritage: A complicated relationship. *Journal of Travel Research, 57*(5), 563–578.

Buhalis, D., & Costa, C. (2005). *Tourism management dynamics*. Butterworth Heinemann.

Bühring-Uhle, C. (1996). *Arbitration and mediation in international business*. Kluwer Law International.

Bull, A. O. (2006). Industrial economics and pricing issues within tourism enterprises and markets. In L. Dwyer & P. Forsyth (Eds.), *International handbook on the economics of tourism* (pp. 138–154). Edward Elgar.

Busser, J., & Shulga, L. (2018). Co-create value: Multidimensional scale and nomological network. *Tourism Management, 65*, 69–86.

Butler, R. (1980). The concept of a tourism area of life cycle of evolution: Implications for management of resources. *Canadian Geographer, 19*(1), 5–12.

Butler, R. (Ed.). (2006a). *The tourism area life cycle* (Vol. 1). Channel View Publications.

Butler, R. (Ed.). (2006b). *The tourism area life cycle* (Vol. 2). Channel View Publications.

Butler, R. (2009). Tourism in the future: Cycles, waves or wheels? *Futures, 41*(6), 346–352.

Butler, R. (2014). Coastal tourist resorts: History, development and models. *ACE: Architecture, City and Environment, 9*(25), 203–228.

Byrne, D. (2002). Industrial culture in a post-industrial world: The case of the North East of England. *City-Analysis of Urban Change, Theory, Action, 6*(3), 279–289.

Calil, C. F. E. (2006). *Posicionamento Social e Desempenho Financeiro: Uma Análise na Indústria Bancária Brasileira a partir de uma Tipologia Proposta*. Unpublished Doctoral Dissertation, PUC-Rio.

Campon-Cerro, A. M., Folgado-Fernendez, J. A., & Manuel, H.-M. J. (2017). Rural destination development based on olive oil tourism: Impact of residents community attachment and quality of life on their support for tourism development. *Sustainability*, *9*(9), 1624.

Campos, A. C., Mendes, J., Oom do Valle, P., & Scott, N. (2016). Co-creation experiences: Attention and memorability. *Journal of Travel & Tourism Marketing*, *33*(9), 309–1336.

Canale, R. R., de Simone, E., di Maio, A., & Parenti, B. (2019). UNESCO World Heritage sites and tourism attractiveness: The case of Italian provinces. *Land Use Policy*, *85*, 114–120.

Capenerhurst, J. (1994). Community leisure and recreation: Theory and practice. In L. Haywood (Ed.), *Community tourism* (pp. 173–192). Butterworth-Heinemann Ltd.

Caust, J., & Vecco, M. (2017). Is UNESCO World Heritage recognition a blessing or burden? Evidence from developing Asian countries. *Journal of Cultural Heritage*, *27*, 1–9.

Cavaco, C. (2009). Turismo sénior: Perfis e práticas. *Cogitur*, *2*(2), 33–64.

Cawley, M., & Gillmor, D. A. (2008). Integrated rural tourism: Concepts and practice. *Annals of Tourism Research*, *35*(2), 316–337.

Cellini, R. (2011). Is UNESCO recognition effective in fostering tourism? A comment on Yang, Lin and Han. *Tourism Management*, *32*(2), 452–454.

Censis. (2016). *Dallo smontaggio della città-fabbrica alla nuova manifattura urbana*. Make in Italy-Censis, Roma. Retrieved from https://www.fondazionergo.it/News/primo-piano/cosa-dice-il-rapporto-del-censis-sulla-manifattura-urbana

César, C. (1985). *Ilha da Madeira, "Paraíso Terrestre" – Sua História, Povo e Mentalidade, Aspecto Social, Económico, Turístico e Cultural*. Editorial Eco do Funchal.

Chaoqun, C. (2011). Research on application of the renewable energy technologies in the development of low-carbon rural tourism. *Energy Procedia*, *5*, 1722–1726.

Chaperon, S., & Bramwell, B. (2013). Dependency and agency in peripheral tourism development. *Annals of Tourism Research*, *40*, 132–154.

Cheer, J., Belhassen, Y., & Kujawa, J. (2017). The search for spirituality in tourism: Toward a conceptual framework for spiritual tourism. *Tourism Management Perspectives*, *24*, 252–256.

Chen, Z. (2018). A qualitative pilot study exploring tourists' pre and post-trip perceptions on the destination image of Macau. *Journal of Travel & Tourism Marketing*, *36*(3), 330–344.

Chen, S. C., & Shoemaker, S. (2014). Age and cohort effects: The American senior tourism market. *Annals of Tourism Research*, *48*, 58–75.

Chernov, V. A., Gutik, T. V., Zatesova, O. M., Kumaneva, I. P., Putrik, Y. S., & Skavronskaya, L. S. (2019). *Historical aspects of tourism and hospitality development*. Pacific State University.

Chernov, V. A., Kurbanova, L. M., & Zatesova, O. M. (2017). The dynamics of tourist flows as the basis for the strategic development of the tourism industry in the Khabarovsk Territory. *Economics and Entrepreneurship*, *10*(2), 936–944.

Chernov, V. A., Skavronskaya, L. S., & Moyle, B. D. (2018). Tourism and hospitality industry and Chinese investment. *Economics and Entrepreneurship*, *10*, 971–974.

Chesbrough, H. W. (2006). *Open innovation models: How to thrive in the new innovation landscape*. Boston, MA: Harvard Business School Press.

Chin, W. (2010). How to write up and report PLS analyses. In E. Vinzi, W. Chin, J. Henseler, & H. Wang (Eds.), *Handbook of partial least square*. Houston: Springer.

Chi, C. G. Q., & Qu, H. (2008). Examining the structural relationships of destination image, tourist satisfaction and destination loyalty, an integrated approach. *Tourism Management*, *29*(4), 624–636.

Chow, C. K. W., & Tsui, W. H. K. (2019). Cross-border tourism: Case study of inbound Russian visitor arrivals to China. *International Journal of Tourism Research*, *21*(5), 693–711.

Cigir, K. (2019). Creating a living lab model for tourism and hospitality businesses to stimulate CSR and sustainability innovations. In G. Passerini & N. Marchettini (Eds.), *Sustainable development and planning 2018, transactions on ecology and the environment* (pp. 569–583). WIT Press.

City of Rovaniemi. (2010). *Rovaniemi Economic Development Programme: Objectives and Action Plan 2011–2016*. Retrieved from http://www.rovaniemi.fi/loader.aspx?id=301ffdf2-399e-46a2-9d9e-32c7c034bcfc

Clark, N. T. (2003). 1. Introduction: Taking entertainment seriously. In N. T. Clark (Ed.), *The city as an entertainment machine* (Research in Urban Policy, Vol. 9, pp. 1–17). Leeds: Emerald Publishing Limited.

Cochran, W. G. (1965). *Técnicas de Amostragem*. Fundo de Cultura.

Coles, T. (2006). Enigma variations? The TALC, marketing models and the descendants of the product life cycle. In R. Butler (Ed.), *The tourism life cycle: Conceptual and theoretical issues* (pp. 49–66). Channel View Publications.

Collins, P., & Cunningham, J. A. (2017). *Creative economies in peripheral regions*. London: Palgrave Macmillan.

Compagnucci, L., Spigarelli, F., Coelho, J., & Duarte, C. (2021). Living labs and user engagement for innovation and sustainability. *Journal of Cleaner Production*, *289*, 125721. doi:10.1016/j.jclepro.2020.125721

Connell, J., Page, S. J., & Meyer, D. (2015). Visitor attractions and events: Responding to seasonality. *Tourism Management*, *46*, 283–298.

Cooper, C. (1990). Resorts in decline: The management response. *Tourism Management*, *11*(1), 63–67.

Cooper, C. (2008). *Tourism: Principles and practice*. Pearson Education.

Cooper, C., Scott, N., & Baggio, R. (2009). Network position and perceptions of destination stakeholder importance. *An International Journal of Tourism and Hospitality Research*, *20*(1), 33–45.

Cordeiro, A. M. (2015). *Tratado da Arbitragem*. Almedina.

Costa, C., Cucca, R., & Torri, R. (2016). Milan: A city lost in the transition from the growth machine paradigm towards a social innovation approach. In T. Brandsen, S. Cattacin, A. Evers, & A. Zimmer (Eds.), *Social innovations in the urban context. Nonprofit and civil society studies* (pp. 125–142). Springer.

Cottam, H., Roe, M., & Challacombe, J. (2007). The impact of transitional changes of maritime passenger transport on Croatian tourism development. *Maritime Policy & Development*, *34*(3), 185–202.

Cox, K. R. (2017). Revisiting 'the city as a growth machine'. *Cambridge Journal of Regions, Economy and Society*, *10*(3), 391–405.

Cranmer, E. E., Tom Dieck, M. C., & Fountoulaki, P. (2020). Exploring the value of augmented reality for tourism. *Tourism Management Perspectives*, *35*, 1–8.

Creswell, J. W., & Plano Clark, V. L. (2007). *Designing and conducting mixed methods research*. SAGE Publications.

Crompton, J. L. (1979). An assessment of the image of Mexico as a vacation destination and the influence of geographical location upon that image. *Journal of Travel Research*, *17*(4), 18–23.

Crotti, M. S. R., & Moretti, S. L. A. (2019). A hospitalidade e a influência do atendimento na percepção de satisfação dos hóspedes em hotéis de luxo no brasil. *Revista Turismo, Visão e Ação*, *21*(3), 309–335.

Cuccia, T., Guccio, C., & Rizzo, I. (2016). The effects of UNESCO World Heritage List inscription on tourism destinations performance in Italian regions. *Economic Modelling*, *53*, 494–508.

Cukier, J. (2002). Tourism and development: Concepts and issues. In R. Sharpley & D. J. Telfer (Eds.), *Tourism employment issues in developing countries: Examples from Indonesia* (pp. 149–164). Channel View Publications.

Cunha, L. (2006). *Economia e Política do Turismo*. Verbo.

Dalonso, Y. S. (2015). *Avaliação de políticas públicas de desenvolvimento turístico das cidades de Gramado (Brasil) e Rovaniemi (Finlândia)*. Doctoral Dissertation, University of Minho, RepositóriUM. Retrieved from https://hdl.handle.net/1822/38439

Dalonso, Y. S., Lourenço, J. M., Remoaldo, P. C., & Panosso Netto, A. (2014). Tourism experience, events and public policies. *Annals of Tourism Research*, *46*, 181–184.

Dan, C. (2015, December 28–29). Positive psychology in tourism [Paper presentation]. In *3rd International Conference on Education, Management, Arts, Economics and Social Science*, Changsha, China. Retrieved from https://www.atlantis-press.com/proceedings/icemaess-15/25848006

Davico, L., Mela, A., & Staricco, L. (2009). *Città Sostenibili. Una Prospettiva Sociologica*. Carocci.

Davies, J. S. (2017). *Urban regime theory*. Wiley & Sons, Ltd.

Declaration on biological diversity and sustainable tourism. (1997). Retrieved from https://www.univeur.org/cuebc/downloads/PDF%20carte/06.%20Berlin%20declaration.PDF

Declaration on tourism and sustainable development in the Mediterranean. (1997). Retrieved from https://www.univeur.org/cuebc/downloads/PDF%20carte/07.%20Calvia%20declaration.PDF

Dedusenko, E. A., & Wagenseil, U. (2020). Sustainability focus in destination management. The case of Russia. *Journal of Environmental Management & Tourism*, *11*(3(43)), 529–537.

D'Elia, A. (2018). *Municipality as a platform: The case of Manifattura Milano.* ServDes2018 – Service Design Proof of Concept – Politecnico di Milano. 18th–20th, June 2018. Linköping University Electronic Press. Retrieved from https://ep.liu.se/ecp/150/059/ecp18150059.pdf

Dell'Agnese, E., & Anzoise, V. (2011). Milan, the unthinking metropolis. *International Planning Studies*, *16*(3), 217–235.

Deng, N., Liu, J., Dai, Y., & Li, H. (2019). Different cultures different photos: A comparison of Shanghai's pictorial destination image. *Tourism Management Perspectives*, *30*, 182–192.

Dente, B., Bobbio, L., & Spada, A. (2005). Government o governance per l'innovazione metropolitana? Milano e Torino a confronto. *Studi Organizzativi*, *2*, 29–47.

Devine, A., & Devine, F. (2011). Planning and developing tourism within a public sector quagmire: Lessons from and for small countries. *Tourism Management*, *32*(6), 1253–1261.

Di Benedetto, A., & Bojanic, D. (1993). Tourism area life cycle extensions. *Annals of Tourism Research*, *20*(3), 557–570.

Dimanche, F., & Andrades, L. (Eds.). (2015). *Tourism in Russia: A management handbook.* Leeds: Emerald Publishing Limited.

Dipartimento per lo Sviluppo e la Coesione Economica. (2013). *Strategia Nazionale per le Aree Interne: Definizione, Obiettivi, Strumenti e Governance.* Retrieved from https://www.agenziacoesione.gov.it/strategia-nazionale-aree-interne/documentazione/

Dipartimento per lo Sviluppo e la Coesione Economica. (2014). *Accordo di Programma Quadro per le Aree Interne.* Retrieved from https://www.agenziacoesione.gov.it/lacoesione/le-politiche-di-coesione-in-italia-2014-2020/strumenti-e-programmi/accordo-di-programma-quadro/

Dipartimento per lo Sviluppo e la Coesione Economica. (2016). *Preliminare di Strategia d'area Grand Paradis.* Retrieved from https://www.agenziacoesione.gov.it/strategia-nazionale-aree-interne/valle-daosta/grand-paradis/

Do Valle, P. O., & Assaker, G. (2016). Using partial least square structural equation modeling in tourism research: A review of past research and recommendations for future applications. *Journal of Travel Research*, *55*(6), 695–708.

Dolomez, Y. (2012). *Renewable energy in Croatia.* Embassy of Belgium-Economic and commercial office.

Domínguez-Quintero, A., González-Rodríguez, M., & Paddison, B. (2018). The mediating role of experience quality on authenticity and satisfaction in the context of cultural-heritage. *Current Issues in Tourism*, *23*(2), 24–260.

Douglas, N. (1997). Applying the life cycle model to Melanesia. *Annals of Tourism Research*, *24*(1), 1–22.

Doxey, G. (1975, September 8–11). A causation theory of visitor-resident irritants, methodology and research inferences – The impact of tourism [Paper presentation]. In *Sixth Annual Conference of the Travel Research Association*, San Diego, CA, USA.

Dredge, D. (2006). Policy networks and the local organization of tourism. *Tourism Management*, *27*, 269–280.

Dunets, A. N., Ivanova, V. N., & Poltarykhin, A. L. (2019). Cross-border tourism cooperation as a basis for sustainable development: A case study. *Entrepreneurship and Sustainability*, *6*(4), 2207–2215.

Dwyer, L., Edwards, D., Mistilis, N., Roman, C., & Scott, N. (2009). Destination and enterprise management for a tourism future. *Tourism Management*, *30*(1), 63–74.

Dwyer, L., & Forsyth, P. (Eds.). (2006). *International handbook on the economics of tourism.* Edward Elgar.

Dwyer, L., Forsyth, P., & Dwyer, W. (2010). *Tourism economics and policy.* Channel View Publications.

Echtner, C. M., & Ritchie, J. (1991). The meaning and measurement of destination image. *Journal of Travel Studies*, *2*(2), 2–12.

Echtner, C. M., & Ritchie, J. (1993). The measurement of destination image: An empirical assessment. *Journal of Travel Research, 31*(4), 3–14.

Edgell Sr, D. L. (2020). *Managing sustainable tourism. A legacy for the future* (3rd ed.). Routledge.

Egger, R., & Neuburger, L. (2020). Augmented, virtual, and mixed reality in tourism. In Z. Xiang, M. Fuchs, U. Gretzel, & W. Höpken (Eds.), *Handbook of e-Tourism* (pp. 1–25). Springer International Publishing.

Egresi, I. O., & Kara, F. (2018). Residents' attitudes to tourists visiting their mosques: A case study from Istanbul, Turkey. *Journal of Tourism and Cultural Change, 16*(1), 1–21.

Ekern, S., Logan, W., Sauge, B., & Sinding-Larsen, A. (2012). Human rights and World Heritage: Preserving our common dignity through rights-based approaches to site management. *International Journal of Heritage Studies, 18*(3), 213–225.

Ek, R., Larsen, J., Hornskov, S. B., & Mansfeldt, O. K. (2008). A dynamic framework of tourist experiences: Space-time and performances in the experience economy. *Scandinavian Journal of Hospitality and Tourism, 8*(2), 122–140.

Elliot, J. (1997). *Tourism: Politics and public sector management.* Routledge.

Elliott, O. V., & Salamon, L. M. (2002). *The tools of government: A guide to the new governance.* Oxford University Press.

Ellyatt, H. (2020, January 30). *Russia closes border with China to prevent spread of the coronavirus.* CNBC. Retrieved from https://www.cnbc.com/2020/01/30/coronavirus-russia-closes-border-with-china-to-prevent-spread.html

Eslami, S., Khalifah, Z., Mardani, A., & Streimikiene, D. (2019). Community attachement, tourism impacts, quality of life and residents' support for sustainable tourism development. *Journal of Travel & Tourism Marketing, 36*(9), 1016–1079.

Esteves, M. (2006). Análise de conteúdo. In J. Á. de Lima & J. A. Pacheco (Eds.), *Fazer investigação: contributos para a elaboração de dissertações e teses* (pp. 105–126). Porto Editora.

European Commission. (2003). *Structure, performance and competitiveness of European tourism and its enterprises.* Enterprise Publications.

European Commission. (2015). *Paris Agreement.* Retrieved from https://ec.europa.eu/clima/policies/strategies/2030_en

European Commission. (2019). *Cooperation between regions and countries.* Retrieved from https://ec.europa.eu/regional_policy/en/policy/cooperation/

European Travel Commission. (2018). *Annual Report 2018.* European Travel Commission.

European Travel Commission. (2019a). *European tourism: Trends & prospects (Q3/2019).* European Travel Commission.

European Travel Commission. (2019b). *Horizon 2022, to boldly go where destination Europe has never gone before.* European Travel Commission.

European Union. (2016). *Overcoming obstacles in border regions – Summary report on the online public consultation.* European Union.

Eusebio, C., Lopes, A., & Carneiro, M. J. (2016). Diverse expenditure patterns of international tourists on Santiago Island – Cape Verde. *Tourism Planning & Development, 14*(3), 389–410.

Eusébio, C., Teixeira, L., & Carneiro, M. J. (Eds.). (2021). *ICT tools and applications for accessible tourism.* Hershey, PA: IGI Global.

Evans, G. (2001). *Cultural planning: An urban renaissance?* Routledge Press.

Evans, G., & Shaw, P. (2004). *The contribution of culture to regeneration in the UK: A review of evidence.* London Metropolitan University.

Falatoonitoosi, E., Schaffer, V., & Kerr, D. (2021). Does sustainable tourism development enhance destination prosperity? *Journal of Hospitality & Tourism Research, 46*(5), 1056–1082.

Falk, M., & Vieru, M. (2019). International tourism demand to Finnish Lapland in the early winter season. *Current Issues in Tourism, 22*(11), 1312–1326.

Farkas, G., & England, P. (Eds.). (1988). *Industries, firms, and jobs: Sociological and economic approach.* Plenum Press.

Farrell, B. H., & Twining-Ward, L. (2004). Reconceptualizing tourism. *Annals of Tourism Research, 31*(2), 274–295.

Fernandes, I. P., & Coelho, M. F. (2002). *Economia do Turismo.* Elsevier.

Ferrari, F. (2007). Genova città di camerieri? In G. Borelli (Ed.), *Capitali del Nord-Ovest* (pp. 79–11). Franco Angeli.

Ferrari, F., & Guala, C. (2017). Mega-events and their legacy: Image and tourism in Genoa, Turin and Milan. *Leisure Studies*, *36*(1), 119–137.

Ferreira, S. D. (2011). Efeito da Visita/Não-Visita na imagem de um destino turístico: uma aproximação metodológica mista. *Turismo em Análise*, *22*(3), 681–699.

Ferrer Tapia, B. (2013). *El contrato de transporte aéreo de pasajeros. Sujetos, estatuto y responsabilidad* (Vol. 12). Librería-Editorial Dykinson.

Filep, S., & Pearce, P. (2013). A blueprint for tourist experience and fulfilment research. In P. Pearce & S. Filep (Eds.), *Tourist experience and fulfilment: Insights from positive psychology* (pp. 223–232). Routledge.

Fitzsimons, G. J., & Morwitz, V. G. (1996). The effect of measuring intent on brand-level purchase behavior. *Journal of Consumer Research*, *23*(1), 1–11.

Flint, J., & Raco, M. (Eds.). (2012). *The future of sustainable cities: Critical reflections*. The Policy Press.

Florida, R. (2002). *The rise of the creative class: And how it's transforming work, leisure and everyday life*. New York, NY: Basic Books.

Fornell, C., & Larcker, D. (1981). Evaluating structural equation models with unobservable variables and measurement error. *Journal of Marketing Research*, *18*(1), 39–50.

Forster, J. (1964). The sociological consequences of tourism. *International Journal of Comparative Sociology*, *5*(2), 217–227.

Fortuna, C. (1999). *Identidades, Percursos, Paisagens Culturais*. Celta Editora.

Frey, B., & Steiner, L. (2011). World Heritage List: Does it make sense? *International Journal of Cultural Policy*, *17*(5), 555–573.

Freytag, T., & Bauder, M. (2018). Bottom-up touristification and urban transformations in Paris. *Tourism Geographies*, *20*(3), 443–460.

Friedman, A. L., & Miles, S. (2002). Developing stakeholder theory. *Journal of Management Studies*, *39*(1), 1–21.

Gabriel, K. R. (1971). The biplot display of matrices with application to principal components analysis. *Biometrika*, *58*, 453–467.

Gabriel, K. R., & Odoroff, C. L. (1990). Biplots in biomedical research. *Statistics in Medicine*, *9*, 469–485.

Galindo, M. P. (1986). Una alternativa de representación simultánea: HJ-Biplot. *Questiio*, *10*(1), 13–23.

Galindo, M. P., & Cuadras, C. M. (1986). *Una extensión del método Biplot y su relación con otras técnicas*. Publicaciones de Bioestadística y Biomatemática. Universidad de Barcelona.

Gallarza, M., Saura, I., & García, H. (2002). Destination image: Towards a conceptual framework. *Annals of Tourism*, *29*(1), 56–78.

Galvani, A., Lew, A. A., & Perez, M. S. (2020). COVID-19 is expanding global consciousness and the sustainability of travel and tourism. *Tourism Geographies*, *22*(3), 567–576.

Gao, Y., & Su, W. (2019). Is the World Heritage just a title for tourism? *Annals of Tourism Research*, *78*, 102748.

Garay, L., & Canoves, G. (2011). Life cycles, stages and tourism history: The Catalonia (Spain) experience. *Annals of Tourism Research*, *38*(2), 651–671.

Garcês, S., Pocinho, M., & Jesus, S. (2017). Positive psychology research as a framework for a new conceptual model in tourism settings. In M. Milcu, M. Stevens, & I. Dahl (Eds.), *Modern research in health, education and social sciences. From evaluation to intervention* (pp. 339–344). Editura Universitară.

Garcês, S., Pocinho, M., & Jesus, S. (2018a). Review of optimism, creativity and spirituality in tourism research. *Tourism and Hospitality Management*, *24*(1), 1–11.

Garcês, S., Pocinho, M., & Jesus, S. (2018b). Development of the tourism wellbeing scale. *Revista de Divulgação Científica – AICA*, *10*, 55–63.

Garcês, S., Pocinho, M., & Jesus, S. N. (2019). Psychological wellbeing as a creative resource for businesses in the tourism industry: A multidisciplinary view. In S. J. Teixeira & J. M. Ferreira

(Eds.), *Multilevel approach to competitiveness in the global tourism industry* (pp. 98–119). Hershey, PA: IGI Global.

Garcês, S., Pocinho, M., & Jesus, S. (2020a). Psychometric analysis of the Tourism Wellbeing Scale (TWS): A multidisciplinary approach. *Anais da Academia Brasileira de Ciências*, *92*(1), 1–12.

Garcês, S., Pocinho, M., & Jesus, S. (2020b). The best tourism island destination in the world and meaningful experiences: A systematic literature review. *Revista Portuguesa de Estudo Regionais*, *53*, 23–34.

Garcia, B. (2004). Cultural policy and urban regeneration in Western European Cities: Lessons from experience, prospects for the future. *Local Economy: The Journal of the Local Economy Policy Unit*, *19*(4), 312–326.

Garcia, B. (2007). Culture-led regeneration cannot resolve structural problems on its own. *Urban Views – EUKN Interviews, 2007–2008*, 15–17.

García-Guzmán, J., Fernández del Carpio, A., de Amescua, A., & Velasco, M. (2013). A process reference model for managing living labs for ICT innovation: A proposal based on ISO/IEC 15504. *Computer Standards & Interfaces*, *36*, 33–41.

García-Hernández, M., de la Calle-Vaquero, M., & Yubero, C. (2017). Cultural heritage and urban tourism: Historic city centres under pressure. *Sustainability*, *9*(8), 1346.

Garelli, S. (1997). *The four fundamental forces of competitiveness*. The World Competitiveness Year Book.

Gartner, W. C. (1986). Temporal influences on image change. *Annals of Tourism Research*, *13*(4), 635–644.

Gartner, W. C. (1993). Image formation process. *Journal of Travel and Tourism Marketing*, *2*(2/3), 191–215.

Garzia, B. (2004). Cultural policy and urban regeneration in Western European cities: Lessons from experience, prospects for the future. *Local Economy: The Journal of the Local Economy Policy Unit*, *19*(4), 312–326.

Gašparović, I., & Gašparović, M. (2019). Determining optimal solar power plant locations based on remote sensing and GIS methods: A case study from Croatia. *Remote Sensing*, *11*(12), 1481.

Gašparović, I., Gašparović, M., & Medak, D. (2018). Determining and analysing solar irradiation based on freely available data: A case study from Croatia. *Environmental Development*, *26*, 55–67.

Genoa Metropolitan City. (2017). *Piano Strategico della Città Metropolitana. Il primo passo per costruire insieme il territorio metropolitano*. Retrieved from https://pianostrategico.cittametropolitana.genova.it/sites/default/files/psm2017/All.1_PSM%20aprile%202017.pdf

Genoa Municipality City. (2017). *Annuario Statistico Edizione 2017. Sistema Statistico Nazionale, Direzione Generale, Struttura di Staff Statistica*. Retrieved from http://statistica.comune.genova.it/pubblicazioni/download/annuario/ANNUARIO_ED_2017/ANNUARIO%202017.pdf

Ger, G., & Belk, R. W. (1996). Cross-cultural differences in materialism. *Journal of Economic Psychology*, *17*(1), 55–77.

Gerosa Bellows, M. (2012). The Buzz in Mexico. National Geographic, January/February 2012. Retrieved from https://www.nationalgeographic.com/travel/article/mexico-honey-traveler

Getz, D. (1986). Models in tourism planning: Towards integration of theory and practice. *Tourism Management*, *7*(1), 21–32.

Getz, D. (1992). Tourism planning and destination lifecycle. *Annals of Tourism Research*, *19*(4), 752–770.

Getz, D. (2007). *Event studies: Theory, research and policy for planned events*. Elsevier/Butterworth-Heinemann.

Getz, D. (2008). Event tourism: Definition, evolution, and research. *Tourism Management*, *29*, 403–428.

Getz, D. (2012). *Event studies: Theory, research and policy for planned events*. London: Routledge.

Getz, D., & Page, S. J. (2016). *Event studies: Theory, research and policy for planned events*. Routledge.

Ghiglione, R., & Matalon, B. (1992). *O Inquérito-Teoria e Prática*. Celta Editora.

Giacometti, A., & Teräs, J. (2019). *Regional economic and social resilience: An exploratory in-depth study in the Nordic countries*. Nordregio. Retrieved from https://www.diva-portal.org/smash/get/diva2:1292899/FULLTEXT01.pdf

Gibson, C. (2010). Geographies of tourism: (Un)ethical encounters. *Progress in Human Geography*, *34*(4), 280–293.

Gilboa, S., Jaffe, E. D., Vianelli, D., Pastore, A., & Herstein, R. (2015). A summated rating scale for measuring city image. *Cities*, *44*(April), 50–59.

Gohary, A., Pourazizi, L., Madani, F., & Chan, E. (2018). Examining Iranian tourists' memorable experiences on destination satisfaction and behavioural intentions. *Current Issues in Tourism*, *23*(2), 131–136.

Gössling, S. (2002). Global environmental consequences of tourism. *Global Environmental Change*, *12*(4), 283–302.

Gössling, S., Hansson, C. B., Hörstmeier, O., & Saggel, S. (2002). Ecological footprint analysis as a tool to assess tourism sustainability. *Ecological Economics*, *43*(2), 199–211.

Gossling, S., Scott, D., & Hall, C. M. (2020). Pandemics, tourism, and global change: A rapid assessment of COVID-19. *Journal of Sustainable Tourism*, *9582*, 1–20.

Gower, J. C., & Hand, D. J. (1996). *Biplots*. London: Chapman & Hall.

Gramado's Tourism Board. (2013). Gramado recebeu mais de 5,7 milhões de visitantes em 2012 [Gramado received over 5,7 million visitors in 2012]. Retrieved from http://www.gramado.rs.gov.br/index.php/Institucionais/Gramado-recebeu-mais-de-57-milhoes-de-visitantes-em-2012.html

Grama, S., & Royer, I. (2013). Competencies for open innovation for entrepreneurs in technological incubators: An empirical test in Romania. *Revue de l'Entrepreneuriat*, *12*(3), 33–58.

Gramsci, A. (1992). Taylorism and Fordism. In A. Gramsci (Ed.), *Prison notebooks* (2011). New York, NY: Columbia University Press.

Graziano, T., & Privitera, D. (2020). Cultural heritage, tourist attractiveness and augmented reality: Insights from Italy. *Journal of Heritage Tourism*, *15*(6), 666–679.

Grossi, E., Sacco, P. L., Tavano Blessi, G., & Cerutti, R. (2011). The impact of culture on the individual subjective well-being of the Italian population: An exploratory study. *Applied Research Quality Life*, *6*, 387–410.

Grossi, E., Tavano Blessi, G., Sacco, P. L., & Buscema, M. (2011). The interaction between culture, health and psychological well-being: Data mining from the Italian culture and well-being project. *Journal of Happiness Studies*, *13*, 129–148.

Guala, C. (2007). *Mega Eventi: Modelli e Storie di Rigenerazione Urbana*. Carocci.

Guala, C. (2015). *Mega Eventi. Immagini e Legacy dalle Olimpiadi Alle Expo*. Carocci.

Guerreiro, M., Mendes, J., Fortuna, C., & Pinto, P. (2020). The dynamic nature of the city image. *Tourism*, *68*(1), 83–99.

Guimont, D., & Lapointe, D. (2016). Empowering local tourism providers to innovate through a living lab process: Does scale matter? *Technology Innovation Management Review*, *6*(11), 18–25.

Guimont, D., Lapointe, D., & Sévigny, A. (2017). Changing collaborative practices in tourism – A living lab case. *Canadian Journal of Regional Science/Revue Canadienne des Sciences Régionales*, *40*(1), 63–69.

Gunn, C. A. (1972). *Vacationscape designing tourist regions*. University of Texas.

Gunn, C. A. (1988). *Tourism planning* (2nd ed.). Taylor & Francis.

Gunn, C. A. (2004). Prospects for tourism planning: Issues and concerns. *The Journal of Tourism Studies*, *15*(1), 3–7.

Gursoy, D., Jurowski, C., & Uysal, M. (2002). Resident attitudes: A structural modeling approach. *Annals of Tourism Research*, *29*(1), 79–105.

Gursoy, D., Ouyang, Z., Nunkoo, R., & Wei, W. (2019). Residents' impact perception of and attitudes towards tourism development: A meta-analysis. *Journal of Hospitality Marketing & Management*, *29*(3), 306–333.

Gursoy, D., & Rutherford, D. (2004). Host attitudes toward tourism: An improved structural model. *Annals of Tourism Research*, *32*(3), 495–516.

Haahti, A., & Yavas, U. (2005). The effect of customer satisfaction with leisure services on behavioral intentions. *Services Marketing Quarterly*, *26*(3), 1–12.

Hadinejad, A., Moyle, B. D., Scott, N., Kralj, A., & Nunkoo, R. (2019). Residents' attitudes to tourism: A review. *Tourism Review*, *74*(2), 150–165.

Hadžić, N., Kozmar, H., & Tomić, M. (2014). Offshore renewable energy in the Adriatic Sea with respect to Croatian 2020 energy strategy. *Renewable and Sustainable Energy Reviews*, *40*, 597–607.

Hair, J., Ringle, C. M., & Sarstedt, M. (2011). PLS-SEM: Indeed a silver bullet. *Journal of Marketing Theory and Practice*, *19*(2), 139–152.

Hair, J., Risher, J., Sarstedt, M., & Ringle, C. (2019). When to use and how to report the results of PLS-SEM. *European Business Review*, *31*(1), 2–24.

Hall, C. M. (2006). Space-time accessibility and the tourist area cycle of evolution: The role of geographies of spatial interaction and mobility in contributing to an improved understanding of tourism. In R. Butler (Ed.), *The tourism life cycle: Conceptual and theoretical issues* (pp. 83–100). Channelview.

Hall, C. M. (2007). Tourism, governance and the (mis-)location of power. In *Tourism, power and space*. Routledge.

Hall, C. M. (2008). Santa Claus, place branding and competition. *Fennia*, *186*(1), 59–67.

Hall, C. M., Harrison, D., Weaver, D., & Wall, G. (2013). Vanishing peripheries: Does tourism consume places? *Tourism Recreation Research*, *38*(1), 71–92.

Hall, C. M., & Jenkins, J. (1995). *Tourism and public policy*. Sage.

Hampton, M. P. (1998). Backpacker tourism and economic development. *Annals of Tourism Research*, *25*(3), 639–660.

Hampton, M. P. (2005). Heritage, local communities and economic development. *Annals of Tourism Research*, *32*(3), 735–759.

Hampton, M. P. (2010). Enclaves and ethnic ties: The local impacts of Singaporean cross-border tourism in Malaysia and Indonesia. *Singapore Journal of Tropical Geography*, *31*(2), 239–253.

Hanafiah, M. H., Jamaluddin, M. R., & Zulkifly, M. I. (2013). Local community attitude and support towards tourism development in Tioman Island, Malaysia. *Procedia-Social and Behavioral Science*, *105*, 792–800.

Hansen, M. H., Hurwitz, W. N., & Madow, W. G. (1953). *Sample survey methods and theory. Methods and applications* (Vol. 1). John Wiley.

Hartman, K. (2016). Destination management in cross-border regions. In H. Wachowiak (Ed.), *Tourism and borders: Contemporary issues, policies and international research* (pp. 109–130). Routledge.

Haukipuro, L., Väinämö, S., & Hyrkäs, P. (2018). Innovation instruments to co-create needs-based solutions in a living lab. *Technology Innovation Management Review*, *8*, 22–35.

Haywood, K. M. (1986). Can the tourist-area life cycle be made operational? *Tourism Management*, *7*(3), 154–167.

Hennessey, S. M., Yun, D., & MacDonald, R. (2009). Influencing the intentions to visit a destination: The case of potential first-time and repeat visitors [Paper presentation]. In *The 2009 Tourism Travel and Research Association International Conference: Advancing Tourism Research Globally*. Retrieved from http://scholarworks.umass.edu/ttra/2009/Presented_Papers/26

Hollands, R. G. (2008). Will the real smart city please stand up? *City*, *12*(3), 303–320.

Hossain, M., Leminen, S., & Westerlund, M. (2019). A systematic review of living lab literature. *Journal of Cleaner Production*, *213*, 976–988.

Hrastnik, B., Franković, B., Mičević, L., Vujčić, R., Horvath, L., & Rajković, D. (2001). Solar and wind energy demonstration zones in the Dalmatian region. *Renewable Energy*, *24*, 501–515.

Hsu, A. Y. C., King, B., Wang, D., & Buhalis, D. (2016). In-destination tour products and the disrupted tourism industry: Progress and prospects. *Information Technology & Tourism*, *16*(4), 413–433.

Hsu, J. Y., & Soong, J. J. (2014). Development of China-Russia relations (1949–2011) limits, opportunities, and economic ties. *Chinese Economy*, *47*(3), 70–87.

Huettermann, M., Thimm, T., Hannich, F., & Bild, C. (2019). Requirements for future digital visitor flow management. *Journal of Tourism Futures*, *5*(3), 241–258.

Hughes, K., & Moscardo, G. (2019). ICT and the future of tourist management. *Journal of Tourism Futures*, *5*(3), 228–240.

Hunt, J. D. (1975). Image as a factor in tourism development. *Journal of Travel Research*, *13*(4), 1–7.

Hwang, K. H. (2014). Finding urban identity through culture-led urban regeneration. *Journal of Urban Management*, *3*(1–2), 67–85.

ICOMOS. (1987). *Charter for the conservation of historic towns and urban area (Washington charter)*. Washington, DC: International Council of Monuments and Site.

Il Sole24Ore. (2019). *Rapporto Ecosistema Urbano*. Retrieved from https://lab24.ilsole24ore.com/ecosistema-urbano/

INE. (2020). *Estatistica do Turismo: Ano 2019*. Instituto Nacional de Estatistica de Cao Verde.

Instituto Brasileiro de Geografia e Estatística. (2013). *Censo Demográfico 2013. [2013 Demographic Census]*. Retrieved from http://www.ibge.gov.br/lojavirtual/fichatecnica.php?codigoproduto=90220

Invest in the Artic. (2009). *Invest in the Artic*. Retrieved from http://www.rovaniemenkehitys.fi/loader.aspx?id=393002c4-4783-40e4-8158-f5098f3c78b4/

Ioannides, D., Nielsen, P., & Billing, P. (2006). Transboundary collaboration in tourism: The case of the Bothnian Arc. *Tourism Geographies*, *8*(2), 122–142.

IPLAM. (1998). *Normas e Diretrizes para intervenções nas áreas de interesse cultural de São Luís: IPLAM/PMSL*. Instituto de Pesquisa e Planejamento do Município de São Luís (IPLAM/PMSL).

Ip, C., Leung, R., & Law, R. (2011). Progress and development of information and communication technologies in hospitality. *International Journal of Contemporary Hospitality Management*, *23*(4), 533–551.

Ires-Arpa. (2017). *Relazione sullo Stato dell'Ambiente in Piemonte 2017*. Ires-Arpa.

Ispra-Snpa. (2019). *Consumo di suolo, dinamiche territoriali e servizi ecosistemici. Report di sistema SNPA, n°8*. Roma. Retrieved from https://issuu.com/snpambiente/docs/rapporto_consumo_di_suolo_20190917

ISTAT. (1973). *5° Industry and services general census (1971)*. Italian National Institute of Statistics-ISTAT Data Warehouse. Retrieved from https://tinyurl.com/z9k8jj6b

ISTAT. (2003). *8° Industry and services general census (2001)*. Italian National Institute of Statistics-ISTAT Data Warehouse. Retrieved from https://www4.istat.it/it/censimento-industria-e-servizi/industria-e-servizi-2001

ISTAT. (2011). *D. S. Popolazione residente—Censimento 2011—Superficie Delle Abitazioni Occupate da Persone Residenti*. ISTAT. Italian National Institute of Statistics. Retrieved from http://dati-censimentopopolazione.istat.it/Index.aspx

ISTAT. (2013). *9° Industry and services census (2011)*. Italian National Institute of Statistics-ISTAT data warehouse. Retrieved from http://dati-censimentoindustriaeservizi.istat.it/Index.aspx

ISTAT. (2020). 19. Turismo. In *Annuario statistico italiano* (pp. 620–648). ISTAT.

Italian Ministry of Economy and Finance. (2017). *Agenzia delle Entrate: Real Estate Observatory 2010–2017*. Retrieved from https://www.agenziaentrate.gov.it/

Ivandić, N., & Marušić, Z. (2007). The hotel industry's import dependence: The case of Croatia. *Tourism*, *55*(3), 297–307.

Ivandić, N., Marušić, Z., Sever, I., Kos, G., & Vrančić, D. (2019). *Tourism satellite account for the republic of Croatia*. Croatian Bureau of Statistics.

Ivandić, N., & Šutalo, I. (2018). The contribution of tourism to the Croatian economy: An IO approach. *Ekonomski Pregled*, *69*, 20–42.

Jackson, M. S., & Inbakaran, R. J. (2006). Evaluating residents' attitudes and intentions to act towards tourism development in regional Victoria, Australia. *International Journal of Tourism Research*, *8*(5), 355–366.

Jafari, J. (1989). Structure of tourism. In S. F. Wit & L. Moutinho (Eds.), *Tourism marketing and management handbook* (pp. 437–442). Prentice-Hall.

Jang, S., Bai, B., Hu, C., & Wu, C.-M. E. (2009). Affect, travel motivation, and travel intention: A senior market. *Journal of Hospitality & Tourism Research*, *33*(1), 51–73.

Jarillo, J. (1993). *Strategic networks: Creating the borderless organisation*. Butterworth-Heinemann.

Jernsand, E. M. (2019). Student living labs as innovation arenas for sustainable tourism. *Tourism Recreation Research*, *44*(3), 337–347.

Jewell, B., Blackman, A., Kuilboer, A., Hyvonen, T., Moscardo, G., & Foster, F. (2004). Factors contributing to successful tourism development in peripheral regions. *Journal of Tourism Studies*, *15*(1), 59–70.

John, P., & Cole, A. (2000). When do institutions, policy sectors, and cities matter? Comparing networks of local policy makers in Britain and France. *Comparative Political Studies*, *33*, 248–268.

Jurowski, C., Uysal, M., & Williams, D. R. (1997). A theoretical analysis of host community resident reactions to tourism. *Journal of Travel Research*, *36*(2), 3–11.

Kamata, H. (2021). Tourist destination residents' attitudes towards tourism during and after the COVID-19 pandemic. *Current Issues in Tourism*, 1–16.

Kantsperger, M., Thees, H., & Eckert, C. (2019). Local participation in tourism development—Roles of non-tourism related residents of the Alpine destination Bad Reichenhall. *Sustainability*, *11*, 6947. doi:10.3390/su11246947

Kazandzhieva, V., & Santana, H. (2019). E-tourism: Definition, development and conceptual framework. *Tourism: An International Interdisciplinary Journal*, *67*(4), 332–350.

Kelly, N., Kelliher, F., Power, J., & Lynch, P. (2020). Unlocking the niche potential of senior tourism through micro-firm owner-manager adaptive capability development. *Tourism Management*, *79*, 1–7.

Kennel, J. (2018). *#overtourism – A real crisis?* James Kennel. Retrieved from https://jameskennell.com/2018/07/03/overtourism-vs-undertourism/

Khanam, D., & Misir, M. A. (2006). *Public private partnership for efficient port operation: A proposal for Chittagong port authority*. Research Gate. Retrieved from https://www.researchgate.net/publication/228833858_Public_Private_Partnership_for_Efficient_Port_Operation_A_Proposal_for_Chittagong_Port_Authority

Khardadiyah, A.-a.-Q. (1889). *Al Masalik wa Al Maamalik*. Brill.

Kim, W., Jun, H. M., Walker, M., & Drane, D. (2015). Evaluating the perceived social impacts of hosting large-scale sport tourism events: Scale development and validation. *Tourism Management*, *48*, 21–32.

Kim, M. J., Lee, C.-K., & Contractor, N. S. (2019). Seniors' usage of mobile social network sites: Applying theories of innovation diffusion and uses and gratifications. *Computers in Human Behavior*, *90*, 60–73.

Kim, H., Oh, C. O., Lee, S., & Lee, S. (2018). Assessing the economic values of World Heritage sites and the effects of perceived authenticity on their values. *International Journal of Tourism Research*, *20*(1), 126–136.

Kim, M. J., & Preis, M. W. (2016). Why seniors use mobile devices: Applying an extended model of goal-directed behavior. *Journal of Travel & Tourism Marketing*, *33*(3), 404–423.

Kim, K., Uysal, M., & Sirgy, M. J. (2013). How does tourism in a community impact the quality of life of community residents? *Tourism Management*, *36*, 527–540.

Kirillova, K., Lehto, X., & Cai, L. (2016). Existential authenticity and anxiety as outcomes: The tourist in the experience economy. *International Journal of Tourism Research*, *19*(1), 13–26.

Kirillova, K., Lehto, X., & Cai, L. (2017). What triggers transformative tourism experiences? *Tourism Recreation Research*, *42*(4), 498–511.

Klimova, B. (2017). Senior tourism and information and communication technologies. In J. J. Park, S.-C. Chen, & K.-K. Raymond Choo (Eds.), *Advanced multimedia and ubiquitous engineering* (pp. 440–445). Springer.

Kock, F., Norfelt, A., Josiassen, A., Assaf, A. G., & Tsionas, M. G. (2020). Understanding the COVID-19 tourist psyche: The evolutionary tourism paradigm. *Annals of Tourism Research*, *85*(August), 103053.

Koens, K., Postma, A., & Papp, B. (2018). Is overtourism overused? Understanding the impact of tourism in a city context. *Journal of Sustainable Tourism*, *10*, 4384.

Koens, K., & Thomas, R. (2016). "You know that's a rip-off": Policies and practices surrounding micro-enterprises and poverty alleviation in South African township tourism. *Journal of Sustainable Tourism*, *24*, 1641–1654.

Kong, W. H., du Cros, H., & Ong, C. E. (2015). Tourism destination image development: A lesson from Macau. *International Journal of Tourism Cities*, *1*(4), 229–316.

Kostakis, I., & Sardinaou, E. (2012). Which factors affect the willingness of tourists to pay for renewable energy? *Renewable Energy*, *38*, 169–172.

Kozak, M., & Buhalis, D. (2019). Cross–border tourism destination marketing: Prerequisites and critical success factors. *Journal of Destination Marketing & Management*, *14*, 1–9.

Koziol, K. (2020). Representing "Russianness" in the Chinese-Russian Borderland. *Eurasian Geography and Economics*, *61*(1), 100–108.

Krajačić, G., Dujić, N., Zmijarević, Z., Vad Mathiesen, B., Anić Vučinić, A., & Carvahlo, M. (2011). Planning for a 100% independent energy system based on smart energy storage for integration of renewable and CO2 emissions reduction. *Applied Thermal Engineering*, *31*(13), 2073–2083.

Krugman, P. (1994). Competitiveness: A danger of obsession. *Foreign Affairs*, *3*(1), 28–44.

Krugman, P. (1997). *Competitiveness: An international economics reader*. Council on Foreign Relations.

Lacey, G. T., Weiler, B., & Peel, V. (2014). Revealing hidden attractions in a rural Kenyan periphery. *Tourism Geographies*, *16*(2), 174–189.

Lai, k., & Li, X. R. (2015). Tourism destination image: Conceptual problems and definitional solutions. *Journal of Travel Research*, *55*(8), 1–16.

Landry, C. (2000). *The creative city: A toolkit for urban innovators*. Earthscan.

Landry, C. (2018). Charles Landry about creative city, creativity, Lviv and modern opportunities. [Video]. Retrieved from https://youtu.be/X_dqdEp0MS0

Lanzarote Charter or Charter for Sustainable Tourism. (1995). Retrieved from https://www.univeur.org/cuebc/downloads/PDF%20carte/03.%20Charter%20for%20sustainable%20tourism.PDF

Lappi. (2008). *Lapland – Vitaly for life: An abstract of Lapland tourism strategy 2007–2010*. The Regional Council of Lapland 2007. Retrieved from http://www.lapinliitto.fi/c/document_library/get_file?folderId=21330&name=DLFE-659.pdf

Lappi. (2011a). *Lapin Matkailutilastollinen vuosikirja 2011 [Tourism facts in Lapland – Statistical review 2011]*. Retrieved from http://www.lappi.fi/lapinliitto/julkaisut_ja_tilastot/matkailu

Lappi. (2011b). *Lappi elämänvoimaa – Matkailustrategia 2011–2014 [Lapland – Pure life force near you – Tourism Strategy 2011–2014]*. Retrieved from http://www.lappi.fi/lapinliitto/c/document_library/get_file?folderId=21330&name=DLFE-9293.pdf

Lappi. (2015). *Lapin matkailustrategia 2015–2018 [Lapland tourism strategy 2015–2018]*. Retrieved from http://www.lappi.fi/lapinliitto/c/document_library/get_file?folderId=2265071&name=DLFE-25498.pdf

Lappi. (2019). *Lapin matkailustrategia 2020–2023 [Lapland tourism strategy 2020–2023]*. Retrieved from http://www.lappi.fi/c/document_library/get_file?folderId=16886&name=DLFE-35026.pdf

Lapland Chamber of Commerce. (2019). In T. Rautajoki & V. Lakkapää (Eds.), *Arctic business forum – YearBook 2019*. Lapland Chamber of Commerce.

Latkova, P., & Vogt, C. A. (2012). Residents' attitudes towards existing and future tourism development in rural communities. *Journal of Travel Research*, *51*(1), 50–67.

Law, R., Leung, D., & Chan, I. C. C. (2019). Progression and development of information and communication technology research in hospitality and tourism: A state-of-the-art review. *International Journal of Contemporary Hospitality Management*, *32*(2), 511–534.

Lee, B., & Groves, D. (2014). Seniors: Technology, leisure, and travel. *International Journal of Humanities and Social Science*, *4*(14), 16–36.

Lee, Y. S., Weaver, D. B., & Prebensen, N. K. (2017). Arctic destinations and attractions as evolving peripheral settings for the production and consumption of peak tourism experiences. In Y.-S. Lee, D. Weaver, & N. K. Prebensen (Eds.), *Arctic tourism experiences: Production, consumption and sustainability* (pp. 1–8). CABI.

Leino, H., & Puumala, E. (2021). What can co-creation do for the citizens? Applying co-creation for the promotion of participation in cities. *Environment and Planning C: Politics and Space*, *39*(4), 781–799.

Leminen, S. (2013). Coordination and participation in living lab networks. *Technology Innovation Management Review*, *3*, 5–14.

Leminen, S. (2015). What are living labs? *Technology Innovation Management Review*, *5*(9), 29–35.

Leminen, S., & Westerlund, M. (2012). Towards innovation in living labs networks. *International Journal of Product Development*, *17*(1–2), 43–59.

Leminen, S., & Westerlund, M. (2016). A framework for understanding the different research avenues of living labs. *International Journal of Technology Marketing*, *11*, 399–420.

Leminen, S., Westerlund, M., & Nyström, A. (2012, September). Living labs as open-innovation networks. *Technology Innovation Management Review*, 6–11.

Lepak, D. P., Smith, K. G., & Taylor, M. S. (2007). Value creation and value capture: A multi-level perspective. *Academy of Management Review*, *32*(1), 180–194.

Levitt, T. (1980, January–February). Marketing success through differentiation of anything. *Harvard Business Review*, 83–91.

Lewis, C., Kerr, G., & Burgess, L. (2019). Positioning a destination as fashionable: The destination fashion conditioning framework. *Tourism Management*, *72*(June), 209–219.

Li, J. (2017). Public participation in tourism development: An analysis of community development in underdevelopment areas. *DEStech Transactions on Environment, Energy and Earth Science*. doi:10.12783/dteees/apees2017/7693

Lien, T. T. N. (2015). Measuring destination image: A case study of Hue, Vietnam. *Hue University Journal of Science*, *113*(14), 73–89.

Lievens, B., Schaffers, H., Turkama, P., Ståhlbröst, A., & Ballon, P. (2011). Cross border living labs networks to support SMEs accessing new markets. In P. Cunningham & M. Cunningham (Eds.), *Proceedings of the eChallenges e-2011 Conference: 26/10/2011-28/10/2011*. International Information Management Corporation (IIMC).

Liguria Regional Tourism Observatory. (2020). *Annual Report 2019*. Regione Liguria, ISNART. Retrieved from https://www.ge.camcom.gov.it/it/elementi-homepage/notizie/allegati-e-immagini/200922-rapporto-annuale-2019-liguria.pdf

Lin, Z., Chen, Y., & Filieri, R. (2017). Resident-tourist value co-creation: The role of residents' perceived tourism impacts and life satisfaction. *Tourism Management*, *61*, 436–442.

Lindberg, K., McCool, S., & Stankey, G. (1997). Rethinking carrying capacity. *Annals of Tourism Research*, *24*(2), 461–465.

Lingua, P. (2019). *Economia in calo dopo il crollo del Morandi, ma Genova resiste*. Retrieved from https://telenord.it/economia-in-calo-dopo-il-crollo-del-morandi-ma-genova-resiste

Lloyd, R., & Clark, T. N. (2003). The city as an entertainment machine. *Research in Urban Sociology: Critical Perspectives on Urban Redevelopment*, *6*, 357–378.

Lodigiani, R. (Ed.). (2019). *Milano 2019. L'anima della metropoli. Ambrosianeum fondazione culturale. Rapporto Sulla Città*. Franco Angeli.

Lopes-Guzman, T., Borges, O., Hernandez-Merino, M., & Cerezo, J. M. (2013). Tourism in Cape Verde: An analysis from the perspective of demand. *Tourism Economics*, *19*(3), 675–688.

Lopes, M. J., & Dredge, D. (2018). Cruise tourism shore excursions: Value for destinations? *Tourism Planning & Development*, *15*(6), 633–652.

Losada, N., Alén, E., Domínguez, T., & Nicolau, J. L. (2016). Travel frequency of seniors tourists. *Tourism Management*, *53*, 88–95.

Loureiro, S. M. C., Guerreiro, J., & Ali, F. (2020). 20 years of research on virtual reality and augmented reality in tourism context: A text-mining approach. *Tourism Management*, *77*, 1–21.

Lourenço, J. M., Álvares, D. F., & Danko, C. C. (2011). Monitoring tourism processes for sustainable development. In D. Damianakos, P. Ventura, & N. Zavrides (Eds.), *Minor communities and natural and cultural heritage: An asset or a liability?* (pp. 23–40). New York, NY: McGraw-Hill.

Lourenço, J., & Bardi, T. (2005, October 17–20). Creative commercial city centres [Paper presentation]. In *41st IsoCaRP Congress 2005 – Making Spaces for the Creative Economy*, Bilbao, Spain.

Lucatelli, S. (2016). Strategia Nazionale per le Aree Interne: un punto a due anni dal lancio della Strategia. *Agriregionieuropa*, *12*(45), 4–10.

Lucatelli, S., & Carlucci, D. (2013). Aree Interne: un potenziale per la crescita economica del Paese. *Agriregionieuropa*, *9*(34), 17–20.

Lucatelli, S., & Storti, D. (2019). La strategia nazionale aree interne e lo sviluppo rurale: scelte operate e criticità incontrate in vista del post 2020. *Agriregionieuropa*, *15*(56), 46–53.

Lucena, M. D. S. (1977). *Avaliação de Desempenho*. McGraw-Hill do Brasil.

Lu, L., Chi, C. G., & Liu, Y. (2015). Authenticity, involvement, and image: Evaluating tourist experiences at historic districts. *Tourism Management, 50*, 85–96.

Luebke, A. (2020). *3 pillars to save EU tourism: Safety, sustainability, and data*. Retrieved from https://www.tourism-review.com/eu-tourism-to-stand-on-safety-sustainability-and-data-news11718

Lui, J. C., & Var, T. (1986). Resident attitudes towards tourism impacts in Hawaii. *Annals of Tourism Research, 13*(2), 193–214.

Lundberg, D. (1971). Why tourists travel. *Cornell Hotel and Restaurant Administration Quarterly, 11*(4), 75–81.

Lupp, G., Zingraff-Hamed, A., Huang, J. J., Oen, A., & Pauleit, S. (2021). Living labs: A concept for co-designing nature-based solutions. *Sustainability, 13*(1), 188.

Luque-Martinez, T., Del Barrio-García, S., Ibanez-Zapata, J. A., & Molina, M. A. R. (2007). Modeling a city's image: The case of Granada. *Cities, 24*(5), 335–352.

Luxmoore, M. (2020, September 20). *On Russia's far eastern frontier, vast stretches of free land, but little interest*. RadioFreeEurope/RadioLiberty. Retrieved from https://www.rferl.org/a/on-russia-s-far-eastern-frontier-acres-of-free-land-but-little-interest/30848156.html

Ma, H. (2000). Competitive advantage and firm performance. *Competitiveness Review, 10*(2), 15–32.

Mabrouki, O., Amirat, Y., Chibani, A., Fernandez, M. V., & De La Cruz, M. N. (2010). Context-aware frame – Work for rural living labs. In I. Rahal & R. Zalila-Wenkstern (Eds.), *Proceedings of SEDE 2010: 19th International Conference on Software Engineering and Data Engineering* (pp. 146–151). International Society for Computers and their Applications, San Francisco, California.

Ma, M., & Hassink, R. (2013). An evolutionary perspective on tourism area development. *Annals of Tourism Research, 41*, 89–109.

Mahutga, M. C., Ma, X., Smith, D. A., & Timberlake, M. (2010). Economic globalisation and the structure of the world city system: The case of airline passenger data. *Urban Studies, 47*(9), 1925–1947.

Majdandžić, L. (2010). *Fotonaponski sustavi-Priručnik*. Tehnička škola Ruđera Boškovića.

Mandić, A., & Garbin Praničević, D. (2019). Progress on the role of ICTs in establishing destination appeal: Implications for smart tourism destination development. *Journal of Hospitality and Tourism Technology, 10*(4), 791–813.

Manila declaration on the social impact of tourism. (1997). Retrieved from https://www.univeur.org/cuebc/downloads/PDF%20carte/70%20Manila.PDF

Manila declaration on world tourism. (1980). Retrieved from https://www.univeur.org/cuebc/downloads/PDF%20carte/65.%20Manila.PDF

Mantino, F., & Lucatelli, S. (2016). Le aree interne in Italia: un laboratorio per lo sviluppo locale. *Agriregionieuropa, 45*, 1–3.

Marine-Roig, E., & Ferrer-Rosell, B. (2018). Measuring the gap between projected and perceived destination images of Catalonia using compositional analysis. *Tourism Management, 68*, 236–249.

Marra, E. (1989). *Componenti Culturali della Qualità Urbana* (Vol. I). Etas Libri.

Marra, E., Mela, A., & Zajczyk, F. (2004). Tempi difficili per la città. In G. Amendola (Ed.), *Anni in salita. Speranze e paure degli italiani* (pp. 275–310). Franco Angeli.

Martín-Santana, J. D., Beerlí-Palacio, A., & Nazzareno, P. A. (2017). Antecedents and consequences of destination image gap. *Annals of Tourism Research, 62*, 13–25.

Martinez, M. G. (2014). Co-creation of value by open innovation: Unlocking new sources of competitive advantage. *Agribusiness, 30*(2), 132–147.

Martínez-Garcia, E. (2013). El turista sénior en Europa: Actualidad y futuro. *Revista Iberoamericana de Turismo, 3*(1), 38–56.

Martini, A. (a cura di). (2011). *Officine grandi riparazioni: Fucina di treni e di vite*. Città di Torino: Rivista Museo Torino. Retrieved from https://www.museotorino.it/resources/pdf/magazine/flip/ogr/files/assets/common/downloads/publication.pdf

Martinotti, G. (1993). *Metropoli*. Il Mulino.

Martins, M. M. (2013). Do mar ao ultramar: a transmigração do lioz português para São Luís do Maranhão. *Revista de História da Arte e Arqueologia, 19*, 101–115.

Marujo, N. (2013). O desenvolvimento do turismo na Ilha da Madeira. *Turydes – Revista de Investigación en Turismo y Desarrollo Local*, *6*(15), 1–16.

Matiza, T. (2020). Post-COVID-19 crisis travel behaviour: Towards mitigating the effects of perceived risk. *Journal of Tourism Futures*, *8*(1), 99–108.

Mayo, E. J. (1973). Regional images and regional travel behavior. In *The Fourth Annual Conference Proceedings of the Travel Research Association*, Sun Valley, Idaho.

Mbaiwa, J. E. (2005). The socio-cultural impacts of tourism development in the Okavango Delta, Botswana. *Journal of Tourism and Cultural Change*, *2*(3), 163–185.

McCool, S. F., & Lime, D. W. (2001). Tourism carrying capacity: Tempting fantasy or useful reality? *Journal of Sustainable Tourism*, *9*(5), 372–388.

McCreary, A., Seekamp, E., Davenport, M., & Smith, J. (2019). Exploring qualitative applications of social media data for place-based assessments in destination planning. *Current Issues in Tourism*, *23*(1), 82–98.

McElroy, J. (2006). Small island tourist economies across the life cycle. *Asia Pacific Viewpoint*, *47*(1), 61–77.

McKercher, B. (2001). Cross-border tourism: An empirical study of tourism into the Pearl River Delta. *Pacific Tourism Review*, *5*(1), 33–41.

McKercher, B., & Dolnicar, S. (2022). Are 10,752 journal articles per year too many? *Annals of Tourism Research*, 103398.

McKercher, B., & Wong, D. Y. (2004). Understanding tourism behavior: Examining the combined effects of prior visitation history and destination status. *Journal of Travel Research*, *43*(2), 171–179.

Medeiros, E. (2019). Cross-border transports and cross-border mobility in EU border regions. *Case Studies on Transport Policy*, *7*(1), 1–12.

Medeiros, T., & Mendes, J. (2021). Realidade aumentada e a inovação das experiências turísticas: Uma revisão integrativa. *Revista Ibérica de Sistemas e Tecnologias de Informação*, *46*(1), 413–425. Retrieved from https://www.proquest.com/openview/436c4790cb1ca8a45ff1f23f5a89e2f0/1?pq-origsite=gscholar&cbl=1006393

Medeiros, T., Moniz, A. I., Mendes, J., Silva, O., Tomás, L., Sousa, M., … Ferreira, J. A. (2021a). Mudanças e potencialidades do turismo sénior nos Açores na visão de *stakeholders*. In T. Medeiros (Ed.), *Turismo Sénior no Destino Açores* (pp. 41–66). Projeto de investigação TU-Sénior55+.

Medeiros, T., Moniz, A. I., Mendes, J., Silva, O., Tomás, L., Sousa, M., … Ferreira, J. A. (2021b). Turistas seniores: contributos para a caracterização no destino Açores. In T. Medeiros (Ed.), *Turismo Sénior no Destino Açores* (pp. 17–39). Projeto de investigação TU-Sénior55+.

Medeiros, T., Moniz, A. I., Mendes, J., Silva, O., Tomás, L., Sousa, M., … Santos, C. (2021). Oportunidades e desafios do turismo sénior nos Açores. In T. Medeiros (Ed.), *Turismo Sénior no Destino Açores* (pp. 68–95). Projeto de investigação TU-Sénior55+.

Medeiros, T., Moniz, A., Tomás, L., Silva, O., Furtado, S., Vieira, V., … Ferreira, J. (2019). Motivations and activities of senior tourists: Challenge of sustainability in islands. In M. D. Sánchez-Fernández & M. Ríos-Manríquez (Eds.), *El Sector Organizacional Turístico desde una Visión Internacional* (pp. 119–135). Red de Nuevas Tendencias de Gestión.

Medeiros, T., Silva, O., Furtado, S., Moniz, A. I., Vieira, V., & Tomás, L. (2020). Health perception, travel concerns, and senior tourist satisfaction. In Á. Rocha, A. Abreu, J. V. de Carvalho, D. Liberato, E. A. González, & P. Liberato (Eds.), *Advances in tourism, technology and smart systems* (Vol. 171, pp. 405–414). Springer.

Medina-Muñoz, D., & García-Falcón, J. M. (2000). Successful relationships between hotels and agencies. *Annals of Tourism Research*, *27*(3), 737–762.

Melis, D. (2017). Il turismo come opportunità di rigenerazione sostenibile dei centri storici minori. *ACE: Architecture, City and Environment*, *11*(33), 181–196.

Mendes, J. (2019). As Tecnologias de Informação e Comunicação no quotidiano social da pessoa idosa: Breve revisão narrativa. *Interações: Sociedade e as Novas Modernidades*, *36*, 100–112.

Mendes, J., Boavida-Portugal, I., & Moniz, A. I. (2023). Tecnologias de Informação e Comunicação (TIC) na Gestão Estratégica de Produtos Turísticos: Uma Revisão Sistemática. In B. C. Reis & A. P. Castela (Eds.), *Turismo Digital, uma comunicação para nómadas intermitentes* (pp. 11–34). Livros Horizonte.

Michalena, E., & Tripanagnostopoulos, Y. (2009). Contribution of the solar energy in the sustainable tourism development of the Mediterranean islands. *Renewable Energy*, *35*, 667–673.

Mihajlović, I., & Strugar, I. (2015). Revisiting the impact of ICT to the tourism development. In *Proceedings of the ENTRENOVA – ENTerprise REsearch InNOVAtion Conference*, Kotor, Montenegro, 10–11 September 2015 (pp. 524–532). IRENET – Society for Advancing Innovation and Research in Economy.

Mihalič, T. (2002). Tourism and economic development issues. In R. Sharpley & D. J. Telfer (Eds.), *Tourism and development: Concepts and issues* (pp. 81–111). Channel View Publications.

Mikhailova, E. V. (2015). Border tourism on the Russian – Chinese border. *Journal of Siberian Federal University. Humanities & Social Sciences*, *3*(8), 437–451.

Mikhailova, E., Wu, C. T., & Chubarov, I. (2018). Blagoveshchensk and Heihe: (Un)contested twin cities on the Sino-Russian border? In J. Garrard & E. Mikhailova (Eds.), *Twin cities: Urban communities, borders and relationship over time* (pp. 288–300). Routledge.

Milan Municipality. (2014). *Linee guida milano smart city*. Retrieved from http://economiaelavoro.comune.milano.it/sites/default/files/2019-02/milano%20smart%20city%20-%20linee%20guida.pdf

Milan Municipality. (2020). *Turismo. Nel 2019 sfiorati 11 milioni di visitatori*. Milan Municipality News, Press Office. Retrieved from https://www.comune.milano.it/-/turismo.-nel-2019-sfiorati-11-milioni-di-visitatori#:~:text=La%20sola%20citt%C3%A0%20di%20Milano,una%20notte%20sotto%20la%20Madonnina

Milano, C. (2018). Overtourism, social unrest and tourismphobia. A controversial debate. *PASOS: Revista de Turismo y Patrimonio Cultural*, *16*, 551–564.

Miles, M. B., & Huberman, A. M. (1994). *Qualitative data analysis: An expanded sourcebook*. Thousand Oaks: Sage Publications.

Mill, R. C., & Morrison, A. (1985). *The tourism system*. Prentice-Hall.

Mill, R. C., & Morrison, A. (2007). *The tourist system*. Kendall/Hunt Publishing.

Mimoso, M. J. (2009). *Arbitragem do Comércio Internacional – medidas Provisórias e Cautelares*. Quid Juris.

Ministry of Economic Affairs and Employment of Finland. (2017). *Tourism is a growing source of export revenue and a significant provider of employment in Finland*. Retrieved from http://tem.fi/documents/1410877/2735818/Finnish_Tourism_2016.pdf/7bdb46bb-beb4-4347-b3af-c619745c7d77

Ministry of Economic, Labour and Entrepreneurship. (2019). *Energy strategy of the republic of Croatia*. Zagreb.

Ministry of Tourism of Brazil. (2010). *Destinos Indutores Recebem Plano de Marketing [Inductors destinations receive marketing plan]*. Retrieved from http://www.turismo.gov.br/turismo/noticias/todas_noticias/20100723-2.html

Ministry of Tourism of Brazil. (2015). *Índice de Competitividade do Turismo Nacional: Relatório Brasil 2015 [National Tourism Competitiveness Index: Brazil 2015 Report]*. Retrieved from http://antigo.turismo.gov.br/sites/default/turismo/o_ministerio/publicacoes/Indice_competitividade/2015/BentoGoncalves_RA_2015.pdf

Mintzberg, H. (1973). Strategy-Making in three modes. *California Management Review*, *16*(2), 44–53.

Mintzberg, H. (1978). Patterns in strategy formation. *Management Science*, *24*(9), 934–948.

Mintzberg, H. (1988). Generic strategies: Toward a comprehensive framework. In R. B. Lamb & P. Shivastava (Eds.), *Advances in strategic management*. Stamford, CT: JAI Press.

Mintzberg, H., & Quinn, J. B. (2001). *O Processo da Estratégia* (3rd ed.). Bookman.

Mitchell, J., & Li, S. (2016). Autonomy found: Estimating the local benefit from tourism in SIDS – The case of Cape Verde. *Journal of Policy Research in Tourism, Leisure & Events*, *9*(2).

Moal-Ulvoas, G. (2017). Positive emotions and spirituality in older travellers. *Annals of Tourism Research*, *66*, 151–158.

Molotch, H. (1976). The city as a growth machine: Toward a political economy of place. *American Journal of Sociology*, *82*(2), 309–332.

Molotch, H., & Logan, J. (1987). *Urban fortunes: The political economy of place*. Berkeley, CA: University of California Press.

Moniz, A. I., Medeiros, T., Silva, O., & Ferreira, J. (2020). The effects of senior tourists' characteristics on travel motivation. In Á. Rocha, A. Abreu, J. V. de Carvalho, D. Liberato, E. A. González, & P. Liberato (Eds.), *Advances in tourism, technology and smart systems* (Vol. 171, pp. 415–425). Springer.

Montebelli, M., & Vavassori, M. (2017). The essential traits of tourism today: Values, motivations and trends. *Symphonya, 1*, 27–31.

Morais, E. P., Cunha, C. R., & Gomes, J. P. (2013). The information and communication technologies in tourism degree courses: The reality of Portugal and Spain. *Journal of -Learning and Higher Education*, 1–10.

Moreno-Luna, L., Robina-Ramírez, R., Sánchez, M. S. O., & Castro-Serrano, J. (2021). Tourism and sustainability in times of COVID-19: The case of Spain. *International Journal of Environmental Research and Public Health, 18*(4), 1859–1887.

Morgan, M., Lugosi, P., & Ritchie, B. (2010). *The tourism and leisure experience*. Channel View Publication.

Mota, M., Guerreiro, M., & Pinto, P. (2016). The world heritage brand and tourism: An approach to historic centre of São Luís, Brazil. In C. Henriques, M. C. Moreira, & P. Bittencourt (Eds.), *Tourism and history: World Heritage – Case studies of Ibero-American space* (pp. 287–311). Interdisciplinary Centre of Social Sciences, University of Minho (CICS.NOVA.UMinho).

Moulaert, F., Rodriguez, A., & Swyngedouw, E. (Eds.). (2005). The globalized city: Economic restructuring and social polarization in European cities. *The Professional Geographer, 57*(2), 332–334.

Moura, A. C. de, Gosling, M. D. S., Christino, J. M. M., & Macedo, S. B. (2017). Aceitação e uso da tecnologia para escolha de destinos turísticos por pessoas da terceira idade: Um estudo usando a UTAUT2. *Revista Brasileira de Pesquisa em Turismo, 11*(2), 239–269.

Moyle, B. D., Moyle, C. L., Bec, A., & Scott, N. (2019). The next frontier in tourism emotion research. *Current Issues in Tourism, 22*(12), 1393–1399.

Nadeau, J., Heslop, L., O'Reilly, N., & Luk, P. (2008). Destination in a country image context. *Annals of Tourism Research, 35*(1), 84–106.

Natal Luz. (2014). *Natal Luz de Gramado: Viva este espetáculo [Natal Luz of Gramado: Live this show]*. Retrieved from http://www.natalluzdegramado.com.br/programacao.html

National Tourism Agency. (2021). Bulletin n.10, dec. 2020. Retrieved from https://www.enit.it/wwwenit/images/multimedia/bollettino10/BOLLETTINO-ENIT-N-10_fin.pdf

Nawijn, J., & Biran, A. (2018). Negative emotions in tourism: A meaningful analysis. *Current Issues in Tourism, 22*(19), 2386–2398.

Nelson, B. (2002). *God's country or Devil's playground: The best nature writing from the Big Bend of Texas*. University of Texas Press.

NEST. (2021). *Tourism think tank 2021 – Points of interest*. Retrieved from https://issuu.com/nestportugal/docs/report_think_tank_points_of_interest

Nguyen, T. A., & Shadrin, N. V. (2008). Integrated coastal zone management in Vietnam: Pattern and perspectives. *Journal of Water Resources and Environmental Engineering, 23*, 297–304.

Nicola, M., Alsafi, Z., Sohrabi, C., Kerwan, A., Al-Jabir, A., Iosifidis, C., & Agha, R. (2020). The socio-economic implications of the coronavirus pandemic (COVID-19): A review. *International Journal of Surgery, 78*, 185–193.

Nilsson, J. H., Eskilsson, L., & Ek, R. (2010). Creating cross-border destinations: Interreg programs and regionalisation in the Baltic Sea rea. *Scandinavian Journal of Hospitality and Tourism, 10*(2), 153–172.

Nofre, J., Giordano, E., Eldridge, A., Martins, J. C., & Sequera, J. (2018). Tourism, nightlife and planning: Challenges and opportunities for community liveability in La Barceloneta. *Tourism Geographies, 20*(3), 377–396.

Novy, J. (2018). 'Destination' Berlin revisited. From (new) tourism towards a pentagon of mobility and place consumption. *Tourism Geographies, 20*(3), 418–442.

Nunkoo, R. (2016). Toward a more comprehensive use of social exchange theory to study residents' attitudes to tourism. *Procedia Economics and Finance, 39*(November), 588–596.

Nunkoo, R., & Gursoy, D. (2012). Residents' support for tourism. An identity perspective. *Annals of Tourism Research, 39*(1), 243–268.

Nunkoo, R., & Ramkissoon, H. (2010a). Community perceptions of tourism in small island states. A conceptual framework. *Journal of Policy Research in Tourism, Leisure and Events*, *2*(1), 51–65.

Nunkoo, R., & Ramkissoon, H. (2010b). Small island urban tourism: A residents' perspective. *Current Issues in Tourism*, *13*(1), 37–60.

Nunkoo, R., & Ramkissoon, H. (2011). Developing a community support model for tourism. *Annals of Tourism Research*, *38*(3), 964–988.

Nunkoo, R., & Ramkissoon, H. (2012). Power, trust, and social exchange and community support. *Annals of Tourism Research*, *39*(2), 997–1032.

Nunkoo, R., Smith, S. L. J., & Ramkissoon, H. (2013). Residents' attitudes to tourism: A longitudinal study of 140 articles from 1984 to 2010. *Journal of Sustainable Tourism*, *21*(1), 5–25.

Nunkoo, R., & So, K. K. F. (2016). Residents' support for tourism: Testing alternative structural models. *Journal of Travel Research*, *55*(7), 847–861.

Nuryanti, W. (1996). Heritage and postmodern tourism. *Annals of Tourism Research*, *23*(2), 249–260.

Nuur, C., & Laestadius, S. (2010). Development in peripheral regions: Case studies in Sweden. *European Urban and Regional Studies*, *17*(3), 293–307.

Nyíri, P., & Breidenbach, J. (2008). The Altai Road: Visions of development across the Russian – Chinese border. *Development and Change*, *39*(1), 123–145.

OECD (Organization for Economic Cooperation and Development). (2020). *Tourism trends and policies 2020*. Paris: OECD Publishing.

OECD (Organization for Economic Cooperation and Development)/Eurostat. (2005). *Guidelines for collecting and interpreting innovation data—The Oslo manual* (3rd ed.). Paris: OECD Publishing.

Okech, R. N. (2010). Socio-cultural impacts of tourism on World Heritage sites: Communities, perspectives of Lamu (Kenya) and Zanzibar Islands, Asia Pacific. *Journal of Tourism Research*, *15*(3), 339–351.

Olympic.com. (2014). *Turin enjoys on-going tourism legacy of 2006 Winter Games*. Retrieved from https://olympics.com/ioc/news/turin-enjoys-on-going-tourism-legacy-of-2006-winter-games

Oneto, A. A. D., Ferreira, J. B., Giovannini, C. J., & Da Silva, J. F. (2015). Confiança e Satisfação na Compra de Turismo Online. *Revista Brasileira de Pesquisa em Turismo*, *9*(2), 221–239. doi:10.7784/rbtur.v9i2.738

Oppermann, M. (1998). Destination threshold potential and the law of repeat visitation. *Journal of Travel Research*, *37*(2), 131–137.

Oppewal, H., Huybers, T., & Crouch, G. I. (2015). Tourist destination and experience choice: A choice experimental analysis of decision sequence effects. *Tourism Management*, *48*, 467–476.

Ostrower, F., & Stone, M. M. (2006). Governance: Research trends, gaps, and future prospects. In W. Powell & R. Steinberg (Eds.), *The non-profit sector: A research handbook* (pp. 612–628). Yale University Press.

Owen, D. (2009). *Green metropolis: Why living smaller, living closer, and driving less are the keys to sustainability*. Riverhead.

Padin, C. (2012). A sustainable tourism planning model: Components and relationships. *European Business Review*, *24*(6), 510–518.

Pallasmaa, J. (2011). *Os Olhos da Pele. A Arquitetura e os Sentidos*. Bookman.

Pan, S.-Y., Gao, M., Kim, H., Shah, K. J., Pei, S.-L., & Chiang, P.-C. (2018). Advances and challenges in sustainable tourism toward a green economy. *The Science of the Total Environment*, *635*, 452–469.

Panrotas.com. (2018). *Serra Gaúcha lidera reservas de famílias no Airbnb [Serra Gaúcha leads family bookings on Airbnb]*. Retrieved from https://www.panrotas.com.br/destinos/pesquisas-e-estatisticas/2018/12/serra-gaucha-lidera-reservas-de-familias-no-airbnb-lista_161112.html

Pappalepore, I., Maitland, R., & Smith, A. (2014). Prosuming creative urban areas. Evidence from East London. *Annals of Tourism Research*, *44*, 227–240.

Pasher, E., Herzog, O., Harir, M., Turjeman-Levi, Y., & Zhiqiang, W. (2018). Creating and enabling ecosystems for open innovation: Challenges and how to cope with them. In *Exploring the culture of open innovation: Towards an altruistic model of economy* (pp. 201–219). doi:10.1108/978-1-78743-789-020181008

Patterson, I., & Balderas, A. (2020). Continuing and emerging trends of senior tourism: A review of the literature. *Journal of Population Ageing*, *13*(3), 385–399.

Paulisic, M., Tankovic, A., & Hrvatin, M. (2016). Managing the service concept in creating and innovative tourism product [Paper presentation]. In *23rd Biennial International Congress Tourism & Hospitality Industry 2016: Trends and Challenges*, Opatija, Croatia. https://bib.irb.hr/datoteka/828666.Paulisic_CuicTankovic_Hrvatin.pdf

Pearce, D. G. (2008). A needs-functions model of tourism distribution. *Annals of Tourism Research*, *35*(1), 148–168.

Pegas, F. d. V., Weaver, D., & Castley, G. (2015). Domestic tourism and sustainability in an emerging economy: Brazil's littoral pleasure periphery. *Journal of Sustainable Tourism*, *23*(5), 748–769.

Peltier, P. (2019). Travel megatrends 2019: Undertourism is the new overtourism. *Skift Magazine*. Retrieved from https://skift.com/2019/02/05/travel-megatrends-2019-undertourism-is-the-new-overtourism/. Accessed on April 01, 2021.

Perčić, M., & Franković, B. (2016). Solarna energija u priobalnompodručju [Solar Energy in coastal energy of Croatia – Today and tomorrow]. *RepublikeHrvatske–danas i sutra. Pomorski zbornik*, (1), 223–229.

Perdue, R. R., Long, P. T., & Allen, L. (1987). Rural resident tourism perception and attitudes by community level of tourism. *Journal of Travel Research*, *28*(3), 3–9.

Perdue, R. R., Long, P. T., & Allen, L. (1990). Resident support for tourism development. *Annals of Tourism Research*, *17*(4), 586–599.

Perelygina, M. (2018). Cultural distance: Visa accessibility approach. Regional tourism: Experience, problems and perspectives [Paper presentation]. In *V International Scientific Conference at the Izhevsk Institute of Computer Science*, Izhevsk, Russian Federation.

Perna, F., Custódio, M. J., & Oliveira, V. (2018). Tourism destination competitiveness: An application model for the south of Portugal versus the mediterranean region of Spain: Competitivtour. *Tourism & Management Studies*, *14*(1), 19–29.

Pesonen, J., Komppula, R., & Riihinen, A. (2015). Typology of senior travellers as users of tourism information technology. *Information Technology & Tourism*, *15*(3), 233–252.

Peteraf, M. A., & Barney, J. B. (2003). Unravelling the resource-based tangle. *Managerial and Decision Economics*, *24*(4), 309–323.

Peterson, C., & Seligman, M. E. P. (2004). *Character strengths and virtues: A handbook and classification*. American Psychological Association.

Pierdicca, R., Paolanti, M., & Frontoni, E. (2019). eTourism: ICT and its role for tourism management. *Journal of Hospitality and Tourism Technology*, *10*(1), 90–106.

Pike, S., & Kotsi, F. (2016). Destination image. Using the Repertory Test to identify salient attributes. *Tourism Management Perspectives*, *18*, 68–73.

Pike, S., & Ryan, C. (2004). Destination positioning analysis through a comparison of cognitive, affective, and conative perceptions. *Journal of Travel Research*, *42*(4), 333–342.

Pine, B., & Gilmore, J. (1998). Welcome to the experience economy. *Harvard Business Review*, *76*(4), 97–105. PMID:10181589.

Pine, J., & Gilmore, J. (2019). *The experience economy: Competing for customer, time, attention, and money*. Harvard Business Review Press.

Pinfield, G. (1997). The use of indicators in local sustainable development planning: A response to Jeb Brugmann. *Local Environment*, *2*(2), 185–187.

Pinheiro, L. L. (2005). *Arbitragem Transnacional – A Determinação do Estatuto da Arbitragem*. Almedina.

Pizam, A. (1978). Tourism's impacts: The social costs to the destination community as perceived by its residents. *Journal of Travel Research*, *16*(4), 8–12.

Podmetina, D., Soderquist, K. E., Petraite, M., & Teplov, R. (2018). Developing a competency model for open innovation: From the individual to the organisational level. *Management Decision*, *56*(6), 1306–1335.

Porter, M. E. (1980). Industry structure and competitive strategy: Keys to profitability. *Financial Analysts Journal*, *36*(4), 30–41.

Porter, M. E. (1985). *Competitive advantage: Techniques for analyzing industries and competitors*. The Free Press.

Porter, M. E. (1996, November/December). What is Strategy? *Harvard Business Review*. Retrieved from https://hbr.org/1996/11/what-is-strategy

Porter, M. E. (1998). The competitiveness of locations. In M. Porter (Ed.), *On competition* (pp. 184–195). Harvard Business Review Book Series.

Porter, M. E. (1998, November/December). Clusters and new economics of competition. *Harvard Business Review*, 77–90.

Postma, A. (2013). *"When the tourists flew in": Critical encounters in the development of tourism*. University of Groningen-The Netherlands.

Postma, A., Buda, D.-M., & Gugerell, K. (2017). The future of city tourism. *Journal of Tourism Futures*, *3*(2), 95–101.

Prahalad, C. K., & Ramaswamy, V. (2004). Co-creation experiences: The next practice of value creation. *Journal of Interactive Marketing*, *18*(3), 5–14.

Prayag, G., & Ryan, C. (2012). Antecedents of tourists' loyalty to Mauritius the role and influence of destination image, place attachment, personal involvement, and satisfaction. *Journal of Travel Research*, *51*(3), 342–356.

Pretes, M. (1995). Postmodern tourism: The Santa Claus industry. *Annals of Tourism Research*, *22*(1), 1–15.

Priday, G., & Pedell, S. (2017). Deepening user involvement through living labs. In *Proceedings of the 29th Australian Conference on Human-Computer Interaction*, Brisbane, QLD, Australia, (OzCHI 2017). doi:10.1145/3152771.3156190

Prokkola, E. (2010). Borders in tourism: The transformation of the Swedish – Finnish border landscape. *Current Issues in Tourism*, *13*(3), 223–238.

Pucihar, A., Malešič, A., Lenart, G., & Borštnar, M. K. (2014). User-centered design of a web-based platform for the sustainable development of tourism services in a living lab context. *Smart Organizations and Smart Artifacts*, 251–266.

Pucihar, A., Zajc, I., Sernec, R., & Lenart, G. (2019). Living lab as an ecosystem for development, demonstration and assessment of autonomous mobility solutions. *Sustainability*, *11*(15), 4095. doi:10.3390/su11154095

Puiggròs, E., Tort González, N., & Fondevila, I. G. (2017). Innovación, turismo y TIC: El caso de los museos de Barcelona. *PASOS: Revista de Turismo y Patrimonio Cultural*, *15*(3), 69–632.

Quivy, R., & Campenhoudt, L. V. (1992). *Manual de investigação em ciências sociais*. Gradiva.

Qu, H., Kim, L. H., & Im, H. H. (2011). A Model of destination branding, integrating the concepts of the branding and destination image. *Tourism Management*, *32*(3), 465–476.

Ramkissoon, H. (2020a). COVID-19 place confinement, pro-social, pro-environmental behaviors, and residents' wellbeing: A new conceptual framework. *Frontiers in Psychology*, *11*(September), 1–11.

Ramkissoon, H. (2020b). Perceived social impacts of tourism and quality-of-life: A new conceptual model. *Journal of Sustainable Tourism*, *0*(0), 1–17.

Ramos-Soler, I., Martínez-Sala, A.-M., & Campillo-Alhama, C. (2019). ICT and the sustainability of World Heritage sites. Analysis of senior citizens' use of tourism apps. *Sustainability*, *11*(11), 1–17.

Rania, N., Coppola, I., Martorana, F., & Migliorini, L. (2019). The collapse of the Morandi Bridge in Genoa on 14 August 2018: A collective traumatic event and its emotional impact linked to the place and loss of a symbol. *Sustainability*, *11*, 6822.

Rasoolimanesh, S. M., & Seyfi, S. (2020). Residents' perceptions and attitudes toward tourism development: A perspective article. *Tourism Review*, *76*(1), 51–57.

Reed, R., & Defillipi, R. J. (1990). Casual ambiguity, barriers to imitation, and sustainable competitive advantage. *Academy of Management Review*, *15*(1), 88–102.

Reilly, M. D. (1990). Free elicitation of descriptive adjectives for tourism image assessment. *Journal of Travel Research*, *28*(4), 69–76.

Remoaldo, P., Ribeiro, J. C., Vareiro, L., & Santos, J. F. (2014). Tourists' perceptions of World Heritage destinations: The case of Guimarães (Portugal). *Tourism and Hospitality Research*, *14*(4), 206–218.

Renda, A. I., da Costa Mendes, J., & Oom do Valle, P. (2011). A structural model approach of residents' perception of tourism impacts in their own quality of life: The municipality of Loulé, Algarve. *Tourism & Management Studies*, *2*, 1088–1091.

Ribeiro, M. A., Pinto, P., Silva, J. A., & Woosnam, K. M. (2017). Residents' attitudes and the adoption of pro-tourism behaviours: The case of developing island countries. *Tourism Management*, *61*(August), 523–537.

Ribeiro, M. A., Woosnam, K. M., & Megerhi, H. A. (2020). The political economy of tourism development in Cape Verde. In M. Novelli, E. A. Adu-Ampong, & M. A. Ribeiro (Eds.), *Routledge handbook of tourism in Africa* (pp. 233–250). London: Routledge. doi:10.4324/9781351022545

Richardson, R. J., Peres, J. A. D. S., Wanderley, J., Correia, L., & Peres, M. (2012). *Pesquisa Social: Métodos e Técnicas* (14. reimpr). São Paulo: Atlas.

Ringle, C. M., Sarstedt, M., Mitchell, R., & Gudergan, S. P. (2020). Partial least squares structural equation modeling in HRM research. *International Journal of Human Resource Management*, *31*(12), 1617–16543.

Ringle, C. M., Wende, S., & Becker, J.-M. (2015). *SmartPLS 3. Bonningstedt: SmartPLS*. Retrieved from www.smartpls.com

Rintala-Gardin, T. (2013). *Rovaniemen matkailun kehitys ja matkailun aluetaloudellinen vaikutus [Rovaniemi tourism development and tourism in regional economic impact]*. Retrieved from https://www.rovaniemi.fi/loader.aspx?id=ddd4e3a5-cf1e-4e70-a954-a49c1dd36cb3

Ritchie, B. W. (2004). Chaos, crises and disasters: A strategic approach to crisis management in the tourism industry. *Tourism Management*, *25*(6), 669–683.

Ritchie, B. W., & Jiang, Y. (2019). A review of research on tourism risk, crisis and disaster management: Launching the annals of tourism research curated collection on tourism risk, crisis and disaster management. *Annals of Tourism Research*, *79*, 102812.

Ritchie, J. B., & Crouch, G. I. (1993). *Competitiveness in international tourism: A framework for understanding and analysis*. World Tourism Education and Research Centre, University of Calgary.

Ritchie, J. B., & Crouch, G. I. (2003). *The competitive destination: A sustainable tourism perspective*. CABI Publishing.

Rizzi, P., & Graziano, P. (2017). Regional perspectives on global tourism trends. *Symphonya – Emerging Issues in Management*, *3*, 11–26.

Robinson, R. N. S., Ritchie, B. W., Kralj, A., Solnet, D. J., Baum, T., & Ford, R. C. (2014). An Asia-Pacific core–periphery futures paradox: Divergent worker and tourist mobilities. *Journal of Travel Research*, *53*(6), 805–818.

Roche, M. (1994). Mega-events and urban policy. *Annals of Tourism Research*, *21*(1), 1–19.

Rozos, E., Makropoulos, C., & Maksimović, Č. (2013). Rethinking urban areas: An example of an integrated blue-green approach. *Water Supply*, *13*(6), 1534–1542.

Ruiz-Gómez, L. M., Navío-Marco, J., & Rodríguez-Hevía, L. F. (2018). Dynamics of digital tourism's consumers in the EU. *Information Technology & Tourism*, *20*, 59–81.

Rumelt, D. P. (1984). Towards a strategic theory of the firm. *Alternative theories of the firm*, *2002*(2), 286–300.

Rumelt, R. P. (2003). What in the world is competitive advantage? *Policy Working Paper*, *105*, 1–5.

Runko Luttenberg, L. (2015). The barriers to renewable energy use in Croatia. *Renewable and Sustainable Energy Reviews*, *49*, 646–654.

Russo, A. P., & Scarnato, A. (2018). "Barcelona in common": A new urban regime for the 21st-century tourist city? *Journal of Urban Affairs*, *40*(4), 455–474.

Ryan, J., & Silvanto, S. (2011). "A brand for all the nations", The development of the World Heritage brand in emerging markets. *Marketing Intelligence & Planning*, *29*(3), 305–318.

Ryzhova, N., & Ioffe, G. (2009). Trans-border exchange between Russia and China: The case of Blagoveshchensk and Heihe. *Eurasian Geography and Economics*, *50*(3), 348–364.

Sacco, P. L., & Blessi, T. G. (2005). European culture capitals and local development strategies: Comparing the Genoa 2004 and Lille 2004 cases. *Homo Oeconomicus*, *24*(1), 111–143.

Saccomani, S. (2019). Rigenerazione urbana e periferie, guardando Torino: contraddittorietà e frammentazione. *Archivio di Studi Urbani e Regionali*, *125*(2), 26–46.

SAGIA. (2019). *Invist Saudi Tourism*. Retrieved from https://investsaudi.sa/media/1376/tourism_brochure_19th_of_feb2.pdf

Sainaghi, R., & Baggio, R. (2019). The World Milan Expo 2015: Short-term effects on hospitality firms. In C. Sousa, I. Vaz de Freitas, & J. Marques (Eds.), *Proceedings of the 2nd International Conference on Tourism Research ICTR 2019* (pp. 266–273). University Portucalense.

Sainaghi, R., & Baggio, R. (2021). Destination events, stability, and turning points of development. *Journal of Travel Research*, *60*(1), 172–183.

Salvatore, R. (2015). Between nature and landscape: The role of community towards an active conservation in protected areas. In R. Gambino & A. Peano (Eds.), *Nature, policies and landscape policies* (pp. 223–231). Springer International Publishing.

Sampieri, R. H., Collado, C. H., Lucio, P. B., Murad, F. C., & Garcia, A. G. Q. (2006). *Metodologia de Pesquisa*. New York, NY: McGraw-Hill.

Sánchez-Pantoja, N., Vidal, R., & Pastor, M. C. (2018). Aesthetic impact of solar energy systems. *Renewable and Sustainable Energy Reviews*, *98*, 227–238.

Sangpikul, A. (2010). Psychology of happiness and tourism. In A. Makinen & P. Hájek (Eds.), *Psychology of happiness* (pp. 189–195). Nova Science Publishers.

Santos, E. (2014). Destination mutant brands: An empirical perspective of tourism professional and host community to create Mutant Brand Cape Verde. *Journal of Destination Marketing & Management*, *3*(1), 68–78. doi:10.1016/j.jdmm.2013.12.008

Santos, R. A., & Bertoldi, J. (2012). O turismo na melhor idade: Uma análise macro deste nicho de mercado. *Revista Científica Eletrônica de Turismo*, *16*(1). Retrieved from http://faef.revista.inf.br/imagens_arquivos/arquivos_destaque/hvEr2BrxvFIFIcR_2013-5-23-17-51-40.pdf

Santos, M. J., Radicchi, E., & Zagnoli, P. (2019). Port's role as a determinant of cruise destination socio-economic sustainability. *Sustainability*, *11*, 4542.

Santos, E., & Rodriguez-Campo, L. (2014). Destination branding: A reflective analysis of Brand Cape Verde. *Place Branding and Public Diplomacy*, *10*, 87–102.

Saretzki, A., Wilken, M., & Wohler, K. (2002). *Lernende Tourismusregionen – Vernetzung als Strategischer Erfolgsfaktor Kleiner und Mittlerer Unternehmen*. Münster: Tourismus – Beitrage zu Wissenschaft und Praxis – Band 3 – LIT Verlag.

Sarstedt, M., Ringle, C. M., Smith, D., Ream, R., & Hair, J. F. (2014). Partial least squares structural equation modeling (PLS-SEM): A useful tool for family business researchers. *Journal of Family Business Strategy*, *5*(1), 105–115.

Sassen, S. (1991). *The global city: New York, London, Tokyo*. Princeton, NJ: Princeton University Press.

Sassen, S. (2008). Re-assembling the urban. *Urban Geographies*, *29*(2), 113–126.

Saufi, A., O'Brien, D., & Wilkins, H. (2014). Inhibitors to host community participation in sustainable tourism development in developing countries. *Journal of Sustainable Tourism*, *22*(5), 801–820.

Saufi, A., Reid, S., & Patiar, A. (2017). Understanding the host community's experiences of creating small autochthonous tourism enterprises in Lombok, Indonesia. In I. B. D. Lima & V. T. King (Eds.), *Tourism and ethnodevelopment* (1st ed., pp. 170–187). Routledge.

Sautter, E. T., & Leisen, B. (1999). Managing stakeholders a tourism planning model. *Annals of Tourism Research*, *26*(2), 312–328.

Savic, B. (2016, December 7). *Behind China and Russia's "special relationship"*. The Diplomat Media. Retrieved from http://thediplomat.com/2016/12/behind-chinaand-russias-special-relationship

Saxena, G., & Ilbery, B. (2008). Integrated rural tourism a border case study. *Annals of Tourism Research*, *35*(1), 233–254.

Scheyvens, R. (2003). Tourism in destination communities. In S. Singh, D. J. Timothy, & R. K. Dowling (Eds.), *Local involvement in managing tourism* (pp. 229–243). CABI Publishing.

Schuurman, D., de Marez, L., & Ballon, P. (2013, November). Open innovation processes in living lab innovation systems: Insights from the LeYLab. *Technology Innovation Management Review*, 28–36.

Schuurman, D., de Marez, L., & Ballon, P. (2016). The impact of living lab methodology on open innovation contributions and outcomes. *Technology Innovation Management Review*, *6*, 7–16.

Scott, A. (2000). *The cultural economies of cities*. Sage.

Secretaria Regional da Economia, Turismo e Cultura. (2016). *Estratégia para o Turismo da Madeira 2017-2021*. Retrieved from http://www.madeirapromotionbureau.com/media/1731/estrategia-para-o-turismo-da-madeira.pdf

Seligman, M. (2011). *Flourish*. New York, NY: Free Press.

Seligman, M. (2013). *Building the state of wellbeing: A strategy for South Australia. Adelaide thinker in residence 2012–2013*. SA Department of the Premier and Cabinet.

Serra, J. (2008). As Tecnologias de Informação e Comunicação no Turismo: A emergência do e-tourism. *Economia e Sociologia*, *86*, 33–42.

Shao, J., Bai, H., Shu, S., & Joppe, M. (2020). Planners' perception of using virtual reality technology in tourism planning. *E-Review of Tourism Research*, *17*(5), 685–695.

Sharp, B., & Dawes, J. (2001). What is differentiation and how does it work? *Journal of Marketing Management*, *17*(7), 739–759.

Sharpley, R. (2000). Tourism and sustainable development: Exploring the theoretical divide. *Journal of Sustainable Tourism*, *8*, 1–19.

Sharpley, R., & Telfer, D. J. (2002). Tourism and economic development issues. In R. Sharpley & D. J. Telfer (Eds.), *Tourism and development: Concepts and issues* (pp. 81–111). Channel View Publications.

Shoval, N. (2018). Urban planning and tourism in European cities. *Tourism Geographies*, *20*(3), 371–376.

Sigala, M. (2020). Tourism and COVID-19: Impact and implications for advancing and resetting industry and research. *Journal of Business Research*, *117*, 312–321.

Silva Filho, O. P. (1986). *Arquitetura luso-brasileira no Maranhão*. Belo Horizonte: Formato.

Silva, O., Medeiros, T., Moniz, A. I., Tomás, L., Furtado, S., & Ferreira, J. (2020). Tourists' characteristics, travel motivation and satisfaction. In Á. Rocha, A. Abreu, J. V. de Carvalho, D. Liberato, E. A. González, & P. Liberato (Eds.), *Advances in tourism, technology and smart systems* (Vol. 171, pp. 427–436). Springer.

Silva, O., Medeiros, T., Moniz, A. I., Tomás, L., Mendes, J., & Vieira, V. (2021). Senior tourism activities and interests in the azores as a destination. In J. V. de Carvalho, Á. Rocha, P. Liberato, & A. Peña (Eds.), *Advances in tourism, technology and systems* (Vol. 208, pp. 309–319). Springer.

Silva, O., Medeiros, T., Moniz, A., Tomás, L., Vieira, V., Furtado, S., . . . Ferreira, J. (2018). Escala de Motivações dos Turistas Seniores. In *Proceedings of the 25th APDR Congress: Circular economy, urban metabolism and regional development: Challenges for a sustainable future* (pp. 368–374). APDR.

Silveira, L., Santos, N., & Perna, F. (2018). Yachts Passing by the West Coast of Portugal – What to due to Make the Marina and the Destination of Figueira da Foz a Nautical Tourism Reference? *Scientific Journal of Maritime Research*, *32*, 182–190.

Singh, S. (2009). Spirituality and tourism: An anthropologist's view. *Tourism Recreation Research*, *34*(2), 143–155.

Skavronskaya, L., Moyle, B., Scott, N., & Kralj, A. (2020). The psychology of novelty in memorable tourism experiences. *Current Issues in Tourism*, *23*(21), 2683–2698.

SMAU. (2014). *Torino SMILE: Smart mobility, inclusion, life & health, energy*. Città di Torino. Retrieved from https://media.smau.it/x-exhibition/upload/multimedia/pdf/2014/05/08/SMAU_Torino_14_SMART_CITY_citta_torino.pdf

Smith, M., & Diekmann, A. (2017). Tourism and wellbeing. *Annals of Tourism Research*, *66*, 1–13.

Smith, M. K., Egedy, T., Csizmady, A., Jancsik, A., Olt, G., & Michalkó, G. (2018). Non-planning and tourism consumption in Budapest's inner city. *Tourism Geographies*, *20*(3), 524–548.

Spirou, C. (2010). *Urban tourism and urban change: Cities in a global economy*. Routledge.

Srinivasan, M., Velu, A., & Mahubabu, B. (2019). Potential environmental impacts of solar energy technologies. *International Journal of Science and Research*, *8*(5), 792–795.

Stabler, M. J. (1995). The image of destination regions, theoretical and empirical aspects. In B. Goodall & G. Ashworth (Eds.), *Marketing in the tourism industry, the promotion of destination regions* (pp. 133–161). Routledge.

Stankov, U., & Gretzel, U. (2020). Tourism 4.0 technologies and tourist experiences: A human-centered design perspective. *Information Technology & Tourism*, *22*(3), 477–488.

Stepchenkova, S., & Morrison, A. M. (2008). Russia's destination image among American pleasure travellers: Revisiting Echtner and Ritchie. *Tourism Management*, *29*, 548–560.

Stern, E., & Krakover, S. (1993). The formation of a composite urban image. *Geographical Analysis*, *25*(2), 130–146.

Sthapit, E. (2018). Antecedents of a memorable hotel experience: Finnish hotels perspective. *Current Issues in Tourism*, *22*(20), 2458–2461.

Stoker, G. (1995). Intergovernmental relations. *Public Administration*, *273*(1), 101–122.

Stokes, R. (2008). Tourism strategy making: Insights to the events tourism domain. *Tourism Management*, *29*(2), 252–262.

Stone, C. N. (1993). Urban regimes and the capacity to govern: A political economy approach. *Journal of Urban Affairs*, *15*(1), 1–28.

Storti, M. (2015). Genova capitale europea della cultura 2004. *Ricerche per la Progettazione del Paesaggio*, *3*(1), 86–89.

Stylidis, D., Belhassen, Y., & Shani, A. (2015). Three tales of a city: Stakeholders' images of Eilat as a tourist destination. *Journal of Travel Research*, *54*(6), 702–716.

Stylidis, D., Biran, A., Sit, J., & Szivas, E. M. (2014). Residents' support for tourism development: The role of residents' place image and perceived tourism impacts. *Tourism Management*, *45*, 260–247.

Stylidis, D., Shani, A., & Belhassen, Y. (2017). Testing an integrated destination image model across residents and tourists. *Tourism Management*, *58*, 184–195.

Sudman, S. (1966). Probability sampling with quotas. *Journal of the American Statistical Association*, *61*(315), 749–771.

Su, M. M., Wall, G., & Xu, K. (2016). Heritage tourism and livelihood sustainability of a resettled rural community: Mount Sanqingshan World Heritage Site, China. *Journal of Sustainable Tourism*, *24*(5), 735–757.

Svensson, B., Nordin, S., & Flagestad, A. (2005). A governance perspective on destination development-exploring partnerships, clusters and innovation systems. *Tourism Review*, *60*(2), 32–37.

Šker, I., & Floričić, T. (2019). Living lab – Intelligent innovation support for tourism development. In *2019 ENTRENOVA Conference Proceedings*. Retrieved from https://www.entrenova.org/conference-proceedings–book-of-abstracts.html

Tagliabue, T. (2012). *Il comune di Venezia e il piano di Marghera: Il quartiere urbano di Marghera*. Università Iuav di Venezia.

Tambovceva, T., Atstaja, D., Tereshina, M., Uvarova, I., & Livina, A. (2020). Sustainability challenges and drivers of cross-border greenway tourism in rural areas. *Sustainability*, *12*(15), 5927.

Tamura, K. (2016). Exploration on strengthening of maritime connectivity. In *Asia-Pacific economic cooperation project TPT 02 2016S*. Tokyo: Asia-Pacific Economic Cooperation Secretariat.

Tapachai, N., & Waryszak, R. (2000). An examination of the role of beneficial image in tourist destination selection. *Journal of Travel Research*, *39*(1), 37–44.

TASS. (2020). *Some sections of Russia's border with China to be closed as of January 31 midnight*. Retrieved from https://tass.com/emergencies/1114597

Telicheva, E. G., & Chernov, V. A. (2020). Sustainable tourism development: Content-review of the tourism market conjuncture during the pandemic. *Scientific, Technical and Economic Cooperation of the Asia-Pacific Countries in the 21st Century*, *2*, 136–140.

Tervo, K., Hall, M., & Saarinen, J. (2013). Christmas tourists' perceptions to climate change in Rovaniemi, Finland. *Tourism Geographies: An International Journal of Tourism Space, Place and Environment*, *15*(2), 292–317.

The Rimini Charter. (2001). Retrieved from http://www.a21italy.it/medias/AA98D81CB4923DF7.pdf

Thees, H., Pechlaner, H., Olbrich, N., & Schuhbert, A. (2020). The living lab as a tool to promote residents' participation in destination governance. *Sustainability*, *12*(3), 1120.

Thompson, C. S. (2004). Host produced rural tourism: Towa's Tokyo antenna shop. *Annals of Tourism Research*, *31*(3), 580–600.

Timothy, D. J., & Butler, R. W. (1995). Cross-boder shopping: A North American perspective. *Annals of Tourism Research*, *22*(1), 16–34.

Timothy, D. J., & Tosun, C. (2003). Tourists' perceptions of the Canada–USA border as a barrier to tourism at the International Peace Garden. *Tourism Management*, *24*(4), 411–421.

Tisdell, C. A. (2020). Economic, social, and political issues raised by COVID-19 pandemic. *Economic Analysis and Policy*, *68*, 17–28.

Tooman, L. A. (1997). Applications of the life-cycle model in tourism. *Annals of Tourism Research*, *24*(1), 214–234.

Tooman, H. (2015). Motivação para viajar e as necessidades do viajante sénior. In A. Albu, C. Chasovschi, H. Müristaja, H. Tooman, & S. Patiar (Eds.), *Manual de Turismo Sénior* (pp. 23–29). SPI – Sociedade Portuguesa de Inovação Consultadoria Empresarial e Fomento da Inovação, S.A. SPI.

Torino Turismo Official. (2020). *Annual Report 2019*. Turismo, Torino e Provincia. Retrieved from https://www.turismotorino.org/sites/default/files/imce/editor/Report_2019_DEF.pdf

Tourism Bill of Rights and Tourist Code. (1985). Retrieved from https://www.aitr.org/wp-content/uploads/2014/04/omt_sofia1985.pdf

Tourism and Sustainable Development (Decision 7/3). (1999). Retrieved from https://www.univeur.org/cuebc/downloads/PDF%20carte/64.%20New%20York.PDF

Tourist Department of Venice. (2017). *Annuario del Tourismo 2017, Assessorato al Turismo, Città di Venezia*. Retrieved from https://www.comune.venezia.it/sites/comune.venezia.it/files/immagini/Turismo/ANNUARIO%202017%20Ver%202.8.1%20cover.pdf

Trace, K., Frielink, B., & Hew, D. (2009). Maritime connectivity in archipelagic Southeast Asia: An overview. *Southeast Asia Working Papers*, *1*, 1–41.

Tribe, J. (1997). The indiscipline of tourism. *Annals of Tourism Research*, *24*(3), 638–657.

Tribe, J. (2010). *Strategy for tourism*. Goodfellow Publishers.

Tribe, J. (2015). *The economics of recreation, leisure and tourism*. Routledge.

Tribe, J., & Airey, D. (2007). *Developments in tourism research*. Elsevier.

Turin Municipality. (2013). *Piano della Mobilità Ciclabile*. Retrieved from http://www.comune.torino.it/trasporti/bm~doc/all1_piano-della-mobilit-ciclabile_emend.pdf

Turismo de Portugal. (2017). *Estratégia Turismo 2027*. Retrieved from https://www.turismodeportugal.pt/SiteCollectionDocuments/estrategia/estrategia-turismo-2027.pdf

Turner, L., & Ash, J. (1975). *The golden hordes: International tourism and the pleasure periphery*. Constable & Robinson.

Ugur, N. G., & Akbiyik, A. (2020). Impacts of COVID-19 on global tourism industry: A cross-regional comparison. *Tourism Management Perspectives*, *36*(September), 100744. doi:10.1016/j.tmp.2020.100744

UNESCO. (2021a). *World Heritage Centre. World Heritage List*. Retrieved from https://whc.unesco.org/en/list/

UNESCO. (2021b). *World Heritage Centre. Historic Center of São Luís*. Retrieved from https://whc.unesco.org/en/list/821

United Nations. (n.d.). *Transforming our World: The 2030 agenda for sustainable development*. Department of Economic and Social Affairs. United Nations. Retrieved from https://sdgs.un.org/2030agenda

UNWTO. (1996). *Agenda 21 for the travel and tourism industry: Towards environmentally sustainable development*. Retrieved from https://www.e-unwto.org/doi/book/10.18111/9789284403714

UNWTO. (2005). *Making tourism more sustainable: A guide for policy makers*. UNWTO.

UNWTO. (2013). *Sustainable tourism for development guidebook: Enhancing capacities for sustainable tourism for development in developing countries*. UNWTO.

UNWTO. (2019). *International tourism highlights*. Retrieved from https://www.e-unwto.org/doi/pdf/10.18111/9789284421152

UNWTO. (2020a). *UNWTO Inclusive Recovery Guide – Sociocultural Impacts of COVID-19* (Issue 4). World Tourism Organization (UNWTO).

UNWTO. (2020b). *World tourism barometer* (Vol. 18, Issue 7). United Nation World Tourism Organization.

UNWTO and UNEP. (2005). *Making tourism more sustainable – A guide for policy makers*. United Nations Environment Programme, Division of Technology, Industry and Economics.

Uysal, M., Sirgy, M. J., Woo, E., & Kim, H. L. (2016). Quality of life (QOL) and well-being research in tourism. *Tourism Management*, *53*, 244–261.

Uysal, M., Siry, M., Woo, E., & Kin, H. (2015). Quality of life (QOL) and well-being research in tourism. *Tourism Management*, *53*, 244–261.

Vada, S., Prentice, C., & Hsiao, A. (2019). The influence of tourism experience and well-being on place attachment. *Journal of Retailing and Consumer Services*, *47*, 322–330.

Van der Borg, J., Costa, P., & Gotti, G. (1996). Tourism in European heritage cities Tourism in European heritage cities. *Annals of Tourism Research*, *23*(2), 306–321.

Vanhove, N. (2005). *The economics of tourism destinations*. Elsevier.

Vanolo, A. (2015). The Fordist city and the creative city: Evolution and resilience in Turin, Italy. *City, Culture and Society*, *6*(3), 69–74.

Vargas-Sánchez, A., Oom do Valle, P., da Costa Mendes, J., & Silva, J. A. (2015). Residents' attitude and level of destination development: An international comparison. *Tourism Management*, *48*(June), 199–210.

Vargas-Sánchez, A., Porras-Bueno, N., & Plaza-Mejía, M. A. (2011). Explaining residents' attitudes to tourism: Is a universal model possible? *Annals of Tourism Research*, *38*(2), 460–480.

Vargas-Sánchez, A., Porras-Bueno, N., & Plaza-Mejía, M. A. (2014). Residents' attitude to tourism and seasonality. *Journal of Travel Research*, *53*(5), 581–596.

Vargas-Sánchez, A., & Riquel-Ligero, F. (2020). Impacts of tourism development and residents' satisfaction: An explanatory model for the case of Huelva, Spain. In P. Pinto & M. Guerreiro (Eds.), *Handbook of research on resident and tourist perspectives on travel destinations* (pp. 41–65). Hershey, PA: IGI Global.

Vargo, S. L., & Lusch, R. F. (2004). Evolving to a new dominant logic for marketing. *Journal of Marketing*, *68*, 1–17.

Varnajot, A. (2019). Digital Rovaniemi: Contemporary and future arctic tourist experiences. *Journal of Tourism Futures*, *6*(1), 6–23.

Vasyukova, E. (2018, November 2). Two countries, one island. *Meduza*. Retrieved from https://meduza.io/en/feature/2018/11/02/two-countries-one-island

Veeckman, C., Schuurman, D., Leminen, S., & Westerlund, M. (2013). Linking living lab characteristics and their outcomes: Towards a conceptual framework. *Technology Innovation Management Review*, *3*, 6–15.

Veršić, Z., Muraj, I., & Binički, M. (2018). Nearly zero-energy buildings in Croatia: Comparison of thermal performance in different climatic regions. *Engineering Power: Bulletin of the Croatian Academy of Engineering*, *13*(4), 16–23.

Vicente, D. M. (2004). *Portugal e a Arbitragem Internacional*. Anuário Janus.

Vicente-Villardón, J. L. (2011). *Multbiplot: Multivariate analysis using biplots*. Universidad de Salamanca, Departamento de Estadística.

Vigolo, V., Siemoni, F., Cassia, F., & Ugolini, M. M. (2018). The effects of travel motivation on satisfaction: The case of older tourists. *International Journal of Business and Social Science*, *9*(2), 19–30.

Villa-Lobos, N., & Vieira, M. B. (2010). *Mais Justiça Administrativa e Fiscal – Arbitragem*. Coimbra Editora.

Visit Rovaniemi. (2019). *Facts in figures*. Retrieved from http://www.visitrovaniemi.fi/media/facts-figures

Voorhees, C. M., Brady, M. K., Calantone, R., & Ramirez, E. (2016). Discriminant validity testing in marketing: An analysis, causes for concerns, and proposed remedies. *Journal of the Academy of Marketing Science*, *44*(1), 119–134.

Wagar, J. A. (1964). The carrying capacity of wild lands for recreation. *Forest Science*, *10*(2), 1–24.

Waligo, V. M., Clarke, J., & Hawkins, R. (2013). Implementing sustainable tourism: A multi-stakeholder involvement management framework. *Tourism Management*, *36*, 342–353.

Wall, G., & Mathieson, A. (2006). *Change, impacts, and opportunities*. Pearson Prentice Hall.

Walton, J. K. (2009). Prospects in tourism history: Evolution, state of play and future developments. *Tourism Management*, *30*(6), 783–793.

Wang, W., Wu, W., Luo, J., & Lu, J. (2017). Information technology usage, motivation, and intention: A case of Chinese urban senior outbound travelers in the Yangtze River Delta region. *Asia Pacific Journal of Tourism Research*, *22*(1), 99–115.

Ward, L. M., & Russell, A. (1981). The psychological representation of molar physical environments. *Journal of Experimental Psychology: General*, *110*(2), 121–152.

Weaver, D. B. (2006). Peripheral tourism, sustainability, ecotourism and a light at the end of the tunnel. *Tourism Recreation Research*, *3*(3), 98–100.

Weaver, D. B. (2017). Core–periphery relationships and the sustainability paradox of small island tourism. *Tourism Recreation Research*, *42*(1), 11–21.

Weaver, D. B. (2018). Creative periphery syndrome? Opportunities for sustainable tourism innovation in Timor-Leste, an early stage destination. *Tourism Recreation Research*, *43*(1), 118–128.

Weaver, D. B., & Lawton, L. (2014). *Tourism management*. Hoboken, NJ: John Wiley & Sons.

Weiermair, K. (2000). Competitiveness. *Encyclopedia of Tourism*, 99–100.

Westerlund, M., & Leminen, S. (2011, October). Managing the challenges of becoming an open innovation company: Experiences from living labs. *Technology Innovation Management Review*, 19–25.

Whitford, M. (2011). A Framework for the development of event public policy: Facilitating regional development. *Tourism Management*, *30*, 674–682.

Więckowski, M., & Cerić, D. (2016). Evolving tourism on the Baltic Sea coast: Perspectives on change in the Polish maritime borderland. *Scandinavian Journal of Hospitality and Tourism*, *16*(1), 98–111.

Willson, G., McIntosh, A., & Zahra, A. (2013). Tourism and spirituality: A phenomenological analysis. *Annals of Tourism Research*, *42*, 150–168.

Windguru. (n.d.). Retrieved from www.windguru.cz

Wood, D., & Gray, B. (1991). Toward a comprehensive theory of collaboration. *Journal of Applied Behavioural Science*, *27*, 139–162.

Woo, E., Kim, H., & Uysal, M. (2015). Life satisfaction and support for tourism development. *Annals of Tourism Research*, *50*, 84–97.

Woo, E., Uysal, M., & Sirgy, M. J. (2018). Tourism impact and stakeholders' quality of life. *Journal of Hospitality & Tourism Research*, *42*(2), 260–286.

World Bank. (2018). *Republic of Cabo Verde adjusting the development model to revive growth and strengthen social inclusion*. World Bank Group.

World Tourism Organization. (1985). *Tourism Bill of Rights and tourist code*. General Assembly of the World Tourism Organization, Sofia.

World Tourism Organization. (1989). *Manila declaration on world tourism*. UNWTO Declarations, Manila.

World Tourism Organization & European Travel Commission (Ed.). (2011). *Handbook on tourism product development*. UNWTO.

Wrenn, D. G., Casazza, J., & Smart, E. (1983). *Urban water development*. Urban Land Institute.

WTTC. (2020a). *Cape Verde, 2020 annual research: Key highlights*. World Travel and Tourism Council.

WTTC. (2020b). *To recovery & beyond: The future of travel and tourism in the wake of COVID-19*. Retrieved from https://wttc.org/Research/To-Recovery-Beyond

Xiao, H., & Smith, S. L. (2010). Professional communication in an applied tourism research community. *Tourism Management*, *31*(3), 402–411.

Yang, F., & Lau, V. (2019). Experiential learning for children at World Heritage sites; the join moderating Effect of brand awareness and generation of Chinese family travelers. *Tourism Management*, *72*, 1–11.

Yang, C. H., Lin, H. L., & Han, C. C. (2010). Analysis of international tourist arrivals in China. The role of World Heritage site. *Tourism Management*, *31*, 827–837.

Yeh, S. S. (2020). Tourism recovery strategy against COVID-19 pandemic. *Tourism Recreation Research*, *46*(2), 188–194.

Yin, R. K. (2003). *Case study research: Design and methods*. SAGE Publications.

Yuan, J. J., Fowler, D. C., Goh, B. K., & Lauderdale, M. K. (2013). Mexican cross-border shoppers' motivations to the USA. *International Journal of Culture, Tourism and Hospitality Research*, *7*(4), 394–410.

Yu, C. P., Cole, S. T., & Chancellor, C. (2018). Resident support for tourism development in rural midwestern (USA) communities: Perceived tourism impacts and community quality of life perspective. *Sustainability*, *10*(3), 1–17.

Zanzotto, A. (1992). In M. Brusatin, W. Dorigo, & G. Morelli (Eds.), *Per Giuseppe Mazzariol*. Viella.

Zelst, B. (2018). Unilateral option arbitration clauses: An unequivocal choice for arbitration under the ECHR? *Maastricht Journal of European and Comparative Law*, *25*(1), 77–86.

Zhang, Y., & Blasco, D. (2021). Destination management amid COVID-19: A case study in La Cerdanya, Spain. *Anatolia*, 1–12.

Zhang, Y., & Xie, P. (2018). Motivational determinates of creative tourism: A case study of Albergue art space in Macau. *Current Issues in Tourism*, *22*(20), 2538–2549.

Zhuang, X., Yao, Y., & Li, J. (2019). Sociocultural impacts of tourism on residents of world cultural heritage sites in China. *Sustainability*, *11*(3), 840–859.

Zorin, I. V., Kaverina, T. P., & Kvartalnov, V. A. (2005). Tourism as a type of activity. *Moscow: Finance and Statistics*, *288*, 140–143.

Zukin, S. (1982). *Loft living: Culture and capital in urban change*. Baltimore, MD: Johns Hopkins University Press.

Zukin, S. (1995). *The cultures of cities*. Blackwell Publishing.

Zukin, S. (1998). Urban lifestyles: Diversity and standardisation in spaces of consumption. *Urban Studies*, *35*(5–6), 825–839.

Zumthor, P. (2006). *Atmosferas: Entornos arquitetónicos*. Gustavo Gili.

INDEX

Printed and bound by CPI Group (UK) Ltd, Croydon, CR0 4YY

13/09/2026

14961026-0005